# NATURAL WONDERS

# OF VERMONT

# NATURAL WONDERS

# OF VERMONT

Barbara Radcliffe & Stillman Rogers

COUNTRY
ROADS

PRESS

COUNTRY ROADS PRESS
Oaks • Pennsylvania

**Natural Wonders of Vermont:**
**A Guide to Parks, Preserves & Wild Places**

Published by Country Roads Press
P.O. Box 838, 2170 West Drive
Oaks, PA 19456

Text design by Studio 3.
Illustrations by Lois Leonard Stock.
Map by Allen Crider.
Typesetting by Typeworks.

ISBN 1-56626-145-7


*Library of Congress Cataloging-in-Publication Data*

Rogers, Barbara Radcliffe.
    Natural wonders of Vermont : a guide to parks, preserves and
wild places / author, Barbara Radcliffe & Stillman Rogers ;
illustrator, Lois Leonard Stock.
        p.    cm.
    Includes index.
    ISBN 1-56626-145-7
    1. Vermont – Guidebooks.   2. Natural history – Vermont –
Guidebooks.   3. Natural areas – Vermont – Guidebooks.
4. Parks – Vermont – Guidebooks.   5. Botanical gardens –
Vermont – Guidebooks.    I. Rogers, Stillman.   II. Title.
F47.3.R643   1995
917.4304′43 – dc20                                        95-15887
                                                              CIP


Printed in the United States of America.
10  9  8  7  6  5  4  3  2  1

*For Carole Belsky and John Norton,
whose hospitality gave us a second home in the
Northeast Kingdom and fond memories
of bogs, laughter, and sunsets on Joe's Pond*

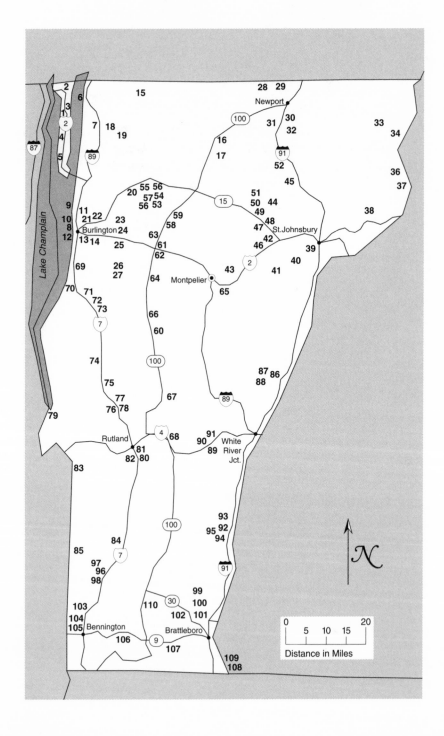

# Contents

## SOUTHERN VERMONT

# Acknowledgments

So many people have taken time to help us in the research necessary to write this book that we cannot mention them all. We can't mention some because we never learned their names—the gentleman out mowing his lawn on a Saturday morning who told us which road we ought to have taken to Shrewsbury, the storekeeper in Pawlet who cautioned us to get to the game supper early, and the charming lady, also in Pawlet, whose number we dialed in error, but who answered our questions anyway. These are Vermonters, some lifelong, some moved here recently, but all people who will take time to help a stranger. It isn't that Vermonters are less busy than others, they simply have priorities that include hospitality and friendliness. However incomplete the list, some require a mention.

Of the many organizations that helped point the way, the Vermont chapter of The Nature Conservancy and the Green Mountain National Forest were of particular help, their staffs eager to send us off to the most interesting and accessible of their treasures.

Our thanks to Bill and Marge Hill of Isle la Motte, Bill

Tecosky and Trish Halloran of Marshfield, Harry Burnham of Island Pond, Michael Bindler of the Phinneas Swan Bed and Breakfast in Montgomery Center (when we weren't even guests!), Maynard Miller of Vernon, Mark Welge of Burke Mountain, Jim Tabor and Jake Elsmeier of Sugarbush, and to Ginger Anderson of the Vermont Department of Forests, Parks and Recreation in Waterbury, for the hours she spent finding us just the right publication to answer our question of the moment. It was Ginger who introduced us to Dale Yerger, who in turn put us in touch with other State Park Naturalists. In addition to Dale, at Little River State Park, we thank the rangers and naturalists, staff, and members of the Youth Conservation Corps at the many other state parks in Vermont, especially Ted Gaine of Groton, Jessica Dillner of North Hero, Wes Fahringer of Underhill, John Stevens at Maidstone, and Mark Guilmette of Brighton. Vermont's parks and forests are among the best managed and best kept we've visited; it is a crowning achievement for a state so small to have set aside so much of its prime lands for public use. The only state park we cannot include in this praise is Wilgus, where we encountered the only rude people we met in the entire time we worked on this book. It's a shame, since the park has a nature trail that local Scouts worked very hard to build.

Innkeepers are, by their choice of career, hospitable sorts. But we have been singularly blessed with hosts who happily spent evenings drawing us maps, calling friends, and otherwise helping us ferret out little-known places. Melanie and Michael Shane took time from a very busy evening at Lilac Inn in Brandon to help us find several interesting places nearby, and Maureen and Bill Russell of the Inn at Montpelier found time for us in the midst of foliage season. Chris Cellars of Grunberg Haus in Duxbury even loaned us an out-of-print book from his personal library. Our thanks to Lee and Beth Davis, Eugene and Jayne Ashley, John and Ann Marie Sherlock, Greg Bohan, and Sal Massaro as well.

And for the generous sharing of information, sources, and ideas, we are forever grateful to our friend and colleague Lisa Rogak. It is not in her character to view this book as competition to her guidebook; rather she sees it as a comfortable companion to her own fine guide to the state.

As always, our family patiently put up with our long absences, dinner-table conversation about bogs and tundra plants, and the general disruption of life that accompanies the writing of a book. To Lura, whose patience seems unbounded; to Julie, whose calls from Japan were most often greeted with our recorded voices; and to Dee, who puts up with us in good humor wherever we may be traveling, our most heartfelt thanks of all.

# Introduction

Vermont's natural wonders come in all sizes and shapes, as well as all degrees of dramatic impact. They are as vast as the thousands of acres of wilderness or as tall as a towering white pine that was alive almost two centuries ago. Others are as small as a well-tended garden or a tiny patch of alpine tundra growing far south of its normal range. Some are as dramatic and loud as the roaring spring runoff rushing through a narrow gorge and over a rock face into a swirling pool far below or as quiet as a bog in the late afternoon when the only sign of activity is a pitcher plant digesting its lunch of fresh mosquito. Some are both quiet and dramatic, as is the sudden sighting of a moose in the twilight, his mouth full of water lilies pulled from a shallow pond beside a Northeast Kingdom road.

The wonder of a place may be in the cry of a loon in the night or the familiar call of *"Oh Sam Peabody"* from a white-throated sparrow in a marsh of an early morning. It may be a taste of tangy blueberries picked beside a mountain trail or the sweet smell of a pine forest on a humid summer afternoon.

Its wonders are colorful—the velvety pink of a fringed

orchis, the orange of a sugar maple in October, the yellow of a brown-eyed Susan in the sunlight, the rich blue of a lake under a clear sky, the reddish purple of a finch, and the gold of a hayfield in the evening sun. Some are spare of colors: the blinding white of a fresh snowfall covering every branch, the deep black of the night sky far from the reflected lights of towns.

Vermont's variety is perhaps its single most appealing asset, for just as you think you know that Vermont is built of marble and granite and slate, you come upon a coral reef complete with fossil shells. The colors change as you watch, like the brilliant pinks and oranges of the sunset over Joe's Pond, fading to deep muted purples and finally to the dark inky blue of night. In the morning, the pond emerges again in a soft layer of white fog that gives way to a pale orange of dawn.

The places we talk about in this book are just as changing, places you can return to over and over, in different seasons, at different times of day, and see anew. The foliage changes color, the sunlight changes direction, the birdsongs change with migrating populations, new flowers bloom and old ones fade, the water level rises abruptly with a thunder shower upstream, the insects take up a concert of chirping as evening approaches, a hickory forest that seemed empty in the spring is filled with scurrying squirrels in the fall. One bog, in North Springfield, not far from our home, we visit almost every time we pass, a tiny place but one that shows us something new each time. We hope our readers will also return to see these places in all lights, all seasons, all moods, and come to know them as we have.

Perhaps the greatest frustration of writing this book has been in having to select which places to include. To describe them all would mean a book the size of the world almanac, certainly too large to carry in a glove compartment or day pack. So we have chosen those that appealed to us most, or that seemed the most illustrative, or that were the easiest to reach. We have also given first choice to those places a traveler would be least likely to hear about elsewhere or find without detailed directions. Which brings us to the subject of signs, or their absence. Unfortunately, much of Vermont's largesse of natural phenomena is hidden to the casual tourist. Few, if any, signs point the way to the many remarkable features of Vermont's natural history – few signs even point the way to the next town, for that matter. So stringent are Vermont's laws against signs that even the major state parks can have only two small roadside directional signs, even though there may be half a dozen roads nearby from which travelers may arrive looking for the park entrance.

Each sign carries a fee, and few private landowners, or public ones either, will pay to direct strangers to the waterfall that happens to be on their land. We do admit that this keeps places uncrowded for the use of local people, and we can't argue with that. It also made our job in writing this book a lot more fun – albeit much more time-consuming – since we had to dig out the information, wheedle the directions, and follow a lot of vague leads.

But we persevered, following the directions of helpful local selectmen, storekeepers, bakers, and residents, until we found most of the places we'd dug up in our research. Some we never did find, such as Jay Branch Gorge. But when we checked with Waterbury, we learned that the people the state had sent out to survey all these geological sites hadn't found it either, although they, too, walked the length of the river. It is, apparently, one of those myths born of a mapping error and perpetuated by travel writers who don't walk the territory.

We never found the Bakersfield Gap, either, although many Bakersfield residents, as puzzled as we were, tried mightily to help us in the search. We leave that one as a challenge to our readers – to the first one who finds it and sends us directions that take us there, goes an autographed copy of the next edition, which will include the name of the pathfinder extraordinaire.

Writing this book has been a series of treasure hunts, often with clues as sparse as an oblique reference in a century-old book. While not all have been successful, all have been enjoyable. If we didn't find what we were looking for, we at least found something interesting—some piece of history, a pleasant trail to hike, a hilltop vista, or a conversation with the owners of the farmhouse in whose yard the road came to an end.

While we can't promise you that some site we describe as accessible will not be fenced and posted in the future or that yet another waterfall will not be ruined by a hydropower station, we can promise you that the places we describe are there and that the way into them is described from our own experience. That's the only way we know how to write a book.

# 1

# Lake Champlain Region

## CHAZAN CORAL REEF

Some 500 million years ago Isle La Motte was a coral reef, and this oldest known reef in the world can be readily seen today, complete with fossilized shells.

*Directions:* Take State 129 from Alburg onto Isle La Motte. Follow the "main road" through the four corners to Quarry Road, where you will see the small stone Historical Society building on the left. Park there. The outcrops of the reef are in the pasture across Quarry Road. The barbed-wire fence is to keep livestock in, but as long as you don't abuse their fence or their hospitality, the owners of the land don't mind your climbing through.

The exposed gray rock of the Chazan Coral Reef lies in low, weather-worn patches in a cow pasture on the old Gilbert Farm. If it isn't enough to know that you are standing on the

1

world's oldest coral reef, and you want to see something to prove it, you can follow from outcrop to outcrop until you find the spiral marks of gastropods, about five inches in diameter, pretty clearly inscribed in the rocks. The exposed surfaces are easy to see, and you can trace them for some distance, spotting gastropod forms on several of them. Kids love it—it's like a treasure hunt to find these little creatures that lived 250 million years before the dinosaurs.

You will, of course, wonder how coral, which is the product of warm tropical seas, came to be in a latitude so far north. The answer lies in continental drift. It is consistent with theories of plate tectonics that this land mass was elsewhere when these corals were alive and forming. This portion of land was at the bottom of a sea much closer to the equator and has drifted northward during the intervening half billion years.

The coral was subject to the same forces as the other rocks that make up the shelf, and with that combination of heat and pressure, became limestone, complete with the fossilized remains of creatures that once lived there.

There is some discussion locally about making some of this area into a designated natural area of some type, with signs and public access. Whatever the future of the reef may be, it is important to protect it. This is *not* a place where specimens can be collected; not only is it on private land and a precious and fragile natural area, but fossils of this variety are not readily removed without destroying them in the process.

## MUD CREEK WILDLIFE MANAGEMENT AREA

The bird-rich marshes at Mud Creek are among the easiest to reach on foot or by wheelchair, since they sit close to a road, with an abandoned railway bed through their center.

**LAKE VERMONT**

Over the course of millions of years, Vermont has been covered by at least four glaciers. Between these glacial periods the western areas of the state, now called the Champlain Valley, filled with water as the glaciers melted. Some 10,000 years ago the Wisconsin Glacier melted and the land at the southern end of the valley rose, causing the water to flow to the north, connecting to the sea through what are now the Richelieu and the Saint Lawrence Rivers. The lake then was much bigger than today's Lake Champlain, extending many miles on either side of the present shores. As the land was relieved of the weight of the glacier, it rose slowly, making the lake smaller until it reached its present size.

*Directions:* East Alburg can be reached from Burlington via US 2 through the islands or I-89 to Swanton. State 78 connects the two, and Mud Creek borders that road on the north, two miles from the end of the causeway/bridge crossing from West Swanton. Two parking areas, the easternmost one at the public boat access, are about .1 mile apart.

An abandoned railbed runs past the boat access, providing a raised walkway from which to view the wetlands with dry feet. Trees along the sides include both silver and red maple and swamp white oak. A trail joins the second, smaller parking area to the railbed, providing a better look at the open waters of the pond, which abounds with duckweed, arum, and water lilies. The railbed offers access suitable for wheelchairs and can be reached from the point where it crosses the road, a few yards west of the boat-access parking area.

Birds are plentiful, especially mallards and blue-winged teal, which nest in the reeds at the edges of the water and feed on the duckweed and sedges. Look sharply for American bittern

**On spring evenings, you can hear *jug-o-rum, jug-o-rum*, the male
bullfrog's resounding mating call booming over ponds**

and green and great blue herons. Other birds, both breeding and
migratory, that you may see here include marsh hawks, black
tern, swamp sparrows, common gallinule, and Virginia rail.

## NORTH HERO TURTLE NESTING SITE
## AND FLOODPLAIN FOREST

Four hundred acres at the northern tip of North Hero Island are
reserved as a nature preserve and protected site for the map
turtle. About a third of that land is below the water level of Lake

Champlain during late winter and spring, creating one of the state's few remaining floodplain forests.

*Directions:* The site is at North Hero State Park. From US 2 north of North Hero, take Lakeview Drive north. Continue north to the main park entrance on the left.

At the east end of North Hero beach is the park's prime draw to naturalists: the nesting grounds of the map turtle, rare in Vermont. During hatching and nesting season, this portion of the beach is roped off to protect the turtles, but early risers who promise to stand still and be quiet can see the newly hatched turtles head for the water very early in the morning. While some of the turtles hatch in the spring, others winter over and hatch about the time of the annual nesting. The park rangers are very knowledgeable about the turtles and can give you details on watching the phenomenon. The turtles should not be touched or bothered in any way during this critical time. The park also has populations of snapping turtles and painted turtles.

In the floodplain forest, also rare in Vermont, ash, red maple, and cottonwood form a nesting habitat for ducks, grouse, and American woodcock. Occupying the grounds of an old farm, the park's fields show the process of reversion to forest. Cutting and burning practices ensure a diversity of habitat that encourages wildlife populations, especially white-tailed deer, which are common.

The campground, Vermont's largest, with 117 sites, is rarely full (July Fourth weekend is the exception), its remote location and lack of trailer hookups discouraging less hardy campers. It opens later than other parks, since half the raised campsites are islands in the flooded forest until the waters recede to summer levels. The shaded, moist forests that make the campsites so pleasant in the summer are a perfect breeding place for mosquitoes, so bring repellent if you camp there. But the camp-sites at North Hero are large, well spaced, and within walking distance of the shingle beach (tiny, flat, worn pebbles—picture a smooth sand that doesn't stick or itch). A bonus is free admission to the sand beach at Knight Point State Park at the other end of the island. Some lean-to sites are graded and ramp equipped for wheelchair access.

Contact the park at RD 1, Box 259, North Hero, VT 05474; 802-372-8727 in summer, 802-879-5674 in winter.

## KNIGHT POINT STATE PARK

One of the smallest of Vermont's state parks, Knight Point protects the longest public stretch of cobbled shoreline on Lake Champlain.

*Directions:* The park is on the west side of US 2 on North Hero Island at the end of the bridge to Grand Isle.

A mile-long walking path circles the point, passing through mature forest with some outstanding large hardwoods. One maple, not far from the beginning of the trail, is one of the largest we have ever seen. Oak, hickory, hornbeam, and cedar create a habitat of various layers for birds and small animals. The trail is level, an easy short walk. Watch overhead for rare osprey, easy to recognize in flight by angular wings that flex at the wrist. Up close, they are distinguished from the gull by black and white face markings.

An area of low bluffs along the shore harbors several endangered plant species, but the land is so fragile that it is fenced to prevent trampling of the rare plants. Those with a serious interest in botany should ask the ranger for permission to visit the area.

A public swimming area and picnic facilities are available onsite. Knight Point State Park, RD 1, Box 21, North Hero, VT 05474; 802-372-8389.

*Of interest in the area:* North Hero is a good base for exploring the islands, and North Hero House, in the center of the lakefront village, offers rooms with a view and meals with a flair. Rates are very reasonable, especially for a waterfront location. Contact North Hero House, PO Box 106, North Hero, VT 05474; 802-372-8237.

## ISLAND SITES FOR ROCK HOUNDS

At least three areas on the Champlain Islands are productive for those who collect mineral specimens and fossils. Vermont law does not allow you to gather fossils or collect minerals in the state parks, although you are welcome to search for them and enjoy them in place.

Several abandoned quarries, which have been fenced off because of the dangers of drowning in the quarry holes, offer good mineral specimens. Serious mineralogists should inquire at the town clerks' offices for the location of the quarries and information on obtaining permission for access to them. Trespass laws are taken seriously in Vermont, so you shouldn't just wander in without permission.

Fossils are found in several places along the lake, and although much of the shore is private land, one site is very easy to reach. From US 2, immediately after it crosses into South Alburg from North Hero, take State 129 and turn right onto the shore road just before the bridge to Isle La Motte. Park north of the public boat access and look in the rocks along the shore.

Quartz and calcite crystals as well as pyrite "suns" are in the rocks used for road fill in building the bridge between North

Hero and Grand Isle. The rocks, located at the northernmost tip of Grand Isle, are best reached from the small access road paralleling US 2, on the east side. Large crystals appear in veins in the black shale, as do pyrite concretions, or "suns." The shingle beach alongside the causeway and the large rocks used as fill both yield interesting specimens.

Trilobite fossils can be found in a road cut in the village of Grand Isle Station, not far from the ferry crossing to New York. State 314 makes a loop on the west side of US 2 and, less than a mile north of the ferry, makes a right-angle curve to the east. Just as the road straightens out after the turn, it passes through the small cut in the bedrock. The black carbon imprints of the chitinous shells appear on both sides of the road, but those on the north are easier to get to (a ditch makes access difficult on the southern face).

## MISSISQUOI NATIONAL WILDLIFE REFUGE

The delta of the Missisquoi River provides an important way stop where migratory birds feed and rest. There is public canoe and trail access to a portion of the 5,600 acres managed by the U.S. Fish and Wildlife Service.

*Directions:* From US 7 or I-89, take State 78 west. The headquarters building for the refuge is about two miles northwest of Swanton on the south side of the road. The refuge lies on both sides of the road, and a boat access is farther along on the north side of the road.

Delta habitats of any size are rare in New England, but Lake Champlain has several. These form a main portion of the Atlantic Flyway, the route followed by birds migrating from their breeding grounds in the Hudson Bay and St. Lawrence

Valley to the southern areas where they winter. During the height of migration, which occurs from mid August through September, over 20,000 ducks may be in the refuge at one time.

Deltas are formed by the sand and silt deposited by rivers as their fast currents, caused by the rapid drop from the mountains and hills, are slowed when their course reaches flat shoreline. Sediments carried off by spring flooding and other high water begin to drop to the riverbed, slowly filling in the shore of the lake. Eventually, these deposits form islands and low marshy lands that extend out into the lake. These wetlands create a habitat ideal for birds.

In addition to the impressive numbers of migratory species, over eighty other varieties are known to nest here, including osprey, wild turkey, Eastern meadowlark, rose-breasted grosbeak, bobolink, and several varieties of owl, woodpecker, warbler, wren, and hawk.

Although the refuge is largely wetland and waterways, an upland area provides habitat for songbirds and small mammals, including star-nosed moles, beavers, deer, muskrats, mink, otters, bobcats, foxes, weasels, several varieties of squirrels, and eight different bats. Along with the more common swamp and woodland flora, wild rice grows in the wetlands, providing food for the waterfowl.

The best way to see the area and its wildlife is by canoe, since 95 percent of it is wetland and because an approach by canoe is quieter and less likely to frighten birds and mammals. Check in at the headquarters to find out what areas are open to visitors and be sure to respect the signs marking the closed areas, such as the rookeries of the great blue heron on Shad Island, which are closed to boaters.

The trip along the Missisquoi River from Swanton, where you can put in just below the dam, to the mouth of the river at Lake Champlain, is about 7.5 miles of flatwater, the last 5.2 inside the refuge. This trip takes you out past the islands of the

**Star-nosed moles sometimes tunnel from land into water,
where they use the feelers that make the stars on their noses to search
for tadpoles and worms on the bottom**

outer delta, including Shad Island, where you must maintain a distance from the shore because of the heron rookery. A marina to the southwest after you enter the lake provides a take-out.

Along with canoe waters, two nature trails give visitors a view of the interaction of the refuge's various inhabitants. The nature trail begins at the headquarters, where a signboard provides maps and interpretive brochures, as well as bird checklists. The nature trails are each about 1.5 miles long on fairly level ground. April through June finds the trail muddy in many places. This is a good time to find abundant animal tracks, but also a good time to wear rubber boots. Insect repellent is helpful in spring and early summer.

Both trails are reached from the same path, which crosses

first a field, then the railway tracks. Beyond this, the Maquam Creek Trail leads to the right, while the Black Creek Trail goes left. Two active beaver lodges are located along the Maquam Creek Trail, one fairly close to its beginning. Several small paths from the trail to the water are made by beavers. At each beaver house is a cleared observation point overlooking the marsh.

Black Creek Trail is bordered by four fern varieties— sensitive, cinnamon, royal, and interrupted—along with spring wildflowers and mosses.

The headquarters building is open daylight hours year-round. For more information, contact Refuge Manager, Missisquoi National Wildlife Refuge, PO Box 163, Swanton, VT; 802-868-4781.

## BURTON ISLAND

A small island in Lake Champlain, once farmed but now preserved as a state park, Burton Island can be explored via walking paths and two interpretive nature trails, either on your own or with the park naturalist through programs offered at its active nature center.

*Directions:* From St. Albans, take State 36 west through the lakeside village of St. Albans Bay to Hathaway Point Road, on the left. Follow the road to its end, into Kill Kare State Park. Ferries to Burton Island run from Kill Kare Park six times a day, from 9:00 A.M. until 6:30 P.M., from the Friday before Memorial Day until Labor Day. The fare is $1.50 each way. Visitors arriving at the island by private boat will find 15 moorings and a 100-slip marina with dockside services.

The 253 acres of Burton Island are characterized by a rocky shoreline, marshes, wooded areas, and open fields overgrowing at the edges to raspberry bramble and sumac. Remnants of its farm, active until the 1960s, include a few foundations, rusted machinery, and an occasional drainage ditch. Other memorabilia of the farm found on the island are displayed at the nature center, where an active naturalist maintains displays of birds' nests, an aquarium, and other changing exhibits.

The Island Farm Nature Trail explores this human history and the ways in which nature has reclaimed the land. An illustrated booklet helps visitors to identify the plants that have overtaken the fields, including vetch and bedstraw, as well as stinging nettle, a plant it advises is "better left alone." The trail booklet is available in large print, braille, and audiocassette; inquire at the nature center.

The North Shore Trail concentrates on the shaping of the island's landscape by glaciers, water, and farmers. The trail begins at the northern end of the campground, and its interpretive brochure helps visitors to read the natural signposts in the landscape, such as the types of plants that grow in the succession process. The trail passes several glacial erratics, boulders carried from elsewhere and dropped by the last retreating glacier, which covered the island with ice one mile thick. The continuing change in landscape features is illustrated by the marsh, which is beginning to fill in with sediment and vegetation and will soon be able to support small shrubs and trees.

Woods Island and Knight Island are also state parklands offering primitive camping by permit only. Although neither has regular ferry service, it is possible to arrange with the Burton Island Ferry for transport to Knight Island. Both of these islands were farmed in the past, and their landscapes are similar to Burton Island. Day visitors arriving by their own boats should be off the islands by 6:00 P.M.

For information on any of these parks, the ferry service, and for permission to camp there, contact Burton Island State

Park, Box 123, St. Albans Bay, VT 05481; in summer 802-524-6353 or in winter 802-879-5674. The telephone number, in summer, for Kill Kare State Park is 802-524-6021.

## WINOOSKI VALLEY PARK DISTRICT

As you explore the natural wonders along the Winooski River between Jericho and its delta on Lake Champlain, you cannot help noticing the large signs with maps of the meandering river and the locations of a string of parks along its banks.

The idea for this park series began in the 1960s when five communities along the lower river agreed to work together in protecting riverside lands as public parks and nature reserves and improving the quality of the river. Two more towns joined some years later, and the result is this chain of parks, the largest of which are in Colchester, Burlington, and Winooski.

We include in the following entries those areas whose prime importance is their natural history, but since the various parks in the valley are spread over such a distance, each is included with other natural attractions in its own area. Canoeists will want to obtain a copy of *Canoe and Natural History Guide* from the district offices or from a local bookstore. This fifty-page booklet describes the canoe route from Cabot to Lake Champlain, with maps, portages, locations of put-ins, and notes on the scenery and difficulty of the water course. To obtain a copy, contact the Winooski River Park District, Ethan Allen Homestead, Burlington, VT 05401; 802-863-5744.

## COLCHESTER BOG

Vermont's only lake-level bog, Colchester Bog is also the only area in the state to support beach heather, a shrub of coastal dunes that spread here when Lake Champlain was connected to the ocean.

*Directions:* From Burlington, follow State 127 north to Porters Point Road in Colchester. Take a left (signs point right to the McCrea Farm at this intersection) and another left at Airport Road. Watch for Airport Park on the right; .4 mile past the entrance is a small pulloff area where you will see the railbed access blocked by stones. If there is not enough space to park there, go back to Airport Park, where plenty of space is available. Although the railbed access is obstructed by large stones, there is room for a wheelchair to pass around these, and the trail is flat and covered by a packed gravel surface.

---

### BEAVERS—WHAT ARE THEY SO BUSY DOING?

While most of these animals live in pairs or very small family groups, they can, and often do, change the entire nature of the area where they decide to settle down. Beavers, of the family *Castor*, are aquatic rodents that choose to live in bogs, marshes, small streams, and along the forested edges of ponds and lakes. A sure sign of their presence is a domed mound of twigs, branches, and mud rising above the surface of a body of water. They live inside these mounds, gaining access by means of a subsurface tunnel. In fact, water serves as their safety net, and if they feel threatened, they give warning to others of their group by sharply beating their large flat tails against the water and submerging in order to swim away from the danger. They feed largely on aquatic plants and the bark of trees such as willow, and when they are building a dam to create a defensive pond they can literally strip a large area of all trees, including some of moderate size. Look for signs of them along the sides of roads and in watery areas. If you see a V-shaped ripple crossing the water, look carefully at the apex, and you'll probably see a small nose cutting through the water.

---

**Dams built by beavers help to prevent flooding and conserve water
for a wide variety of animals and plants**

Once a part of Lake Champlain, the lowlands that comprise
the bog were separated about 8,000 years ago by sand deposits
from the Winooski River. Cut off from its supply of oxygen and
nutrient-rich water from the lake, the lowland began to accumulate peat, which now is as deep as twenty feet in some places.
The railbed, its tracks now removed, provides not only a high
path through the area, but also a unique habitat in itself, where
upland bird species thrive. You may see pileated woodpeckers,
cardinals, and chickadees alongside marsh and pond birds such

as belted kingfishers, mallards, and great blue herons. Flora includes tamarack, American basswood, rhodora, sheep laurel, black raspberries, and a variety of wildflowers, including (in the early spring) hepatica.

Since there is no boardwalk onto the bog surface, the closest visitors can get is the railbed or an observation area reached by a path leaving the road between the railbed and Airport Park. The path leads through a woodlot of American beech, maple, and witch hazel, through ferns to the edge of the water. From the grassy observation area you can see a black tupelo tree to your right, north of its usual range.

For more information about the bog, contact the University of Vermont Environmental Program, 153 South Prospect Street, Burlington, VT 05405; 802-656-4055.

## DELTA PARK

The Winooski River has formed a long sand beach along its northern bank as it enters Lake Champlain. Tree stumps bleached and gray from drifting in the lake have washed ashore and lie strewn on the sand among tufts of dune grass and rare beach peas.

*Directions:* From Burlington, take State 127, following the directions to the Colchester Bog (above) onto Airport Road. Take the second left onto Windermere Way and continue to the public boat access, where there is ample parking. The trail to the shore begins at the end of the road, just past the parking area.

The sand deposited by the river here serves as a foothold for grasses and small shrubs long enough to stabilize the dunes and capture more sand, which in turn creates a base for water-tolerant trees such as willow and cottonwood. The wind off the lake blows the sand, extending the dune area inland. Huge pieces of driftwood, often an entire tree root system and long sections of tree trunk, trap more deposits, and the delta continues to grow.

In midsummer, the path to the shore is alive with tiny frogs, their vivid green and brown markings blending into the leaf cover of the path so that the leaves themselves seem to be jumping from the trail ahead as you walk. The trail forks, the left branch ending at an observation platform, once the abutment for a railroad bridge. Below the bunkerlike structure lies the beach, where snapping turtles breed and nest in the spring. The trees close to the platform are Eastern cottonwoods, moisture-tolerant trees that secure the soil for further growth of the delta.

Behind the beach is the path that forked to the right previously, running parallel to the shore and continuing to the point, the end of the delta. As you stand here on a summer day, picture the scene in February after a storm, when plates of ice, sometimes ten or twelve feet high, lie piled on the shore where the waves and wind have tossed them. The debris frozen into them drops as they melt, further building the shore and depositing new seeds to take hold there.

Behind the beach trail, a swampy lowland provides bird habitat; a platform with a bench, shaded by overhanging trees, overlooks the marsh from beside the entrance trail. You may notice boxes in this area, which are placed to encourage wood ducks to nest there.

Delta Park is managed by the Winooski Valley Park District, Ethan Allen Homestead, Burlington, VT 05401; 802-863-5744.

## McCREA FARM

Lying along a deep curve in the Winooski River, the McCrea Farm's hayfields now provide food and cover for small animals and birds, while its upland fields and forests are filled with wild-flowers in the spring.

*Directions:* From Burlington, follow State 127 north into Colchester, then go right on McCrea Road. There is a sign for the farm on the right side of this intersection. Follow the road until it ends at the beginning of the two nature trails.

The water visible from the road as you drive in to the McCrea Farm is a long, looping cove in the Winooski River, a favorite place for canoe exploration. This cove and the curve in the river provide the farm with considerable water frontage in a relatively small space. The long peninsula between the river and the curving cove almost forms an island. The riverside Loop Trail explores this riverbank and the hayfields.

Wildflowers and vines are plentiful here, including boneset, wild cucumber, vervain, and elder, the berries of which are a particular favorite of songbirds. The trail makes a series of three loops, the first of which leads past a giant maple, whose trunk is easily six feet in diameter. Notice how this and other trees overhanging the river are losing a little more of the bank to which they cling with each spring flood.

The wooden platforms you see on the riverbanks were built there to encourage ospreys to nest. Ospreys, whose wingspread may be seventy inches, are seen at McCrea Farm, but have not yet begun to nest here. They return from South America in April.

For information on the McCrea Farm, contact the Winooski Valley Park District, Ethan Allen Homestead, Burlington, VT 05401; 802-863-5744.

## ETHAN ALLEN HOMESTEAD

The name of Ethan Allen is everywhere in Vermont, from the first place you cross the border in Bennington (site of his most famous battle), to the Burlington area, where he farmed until his death in February 1789. The area surrounding the riverside homestead is now a nature reserve with interpretive trails.

*Directions:* From central Burlington, take State 127 to the exit for "Northern Beaches." Follow the small brown signs to Ethan Allen Homestead. It is also signposted from North Avenue.

Trails lead from the Orientation Center (where a program presents the historical background of Ethan Allen and his times) to various parts of the property. Maps are available at the information board in front of the center, as are interpretive brochures for a self-guided tour of the Wetlands Nature Trail.

Along with helping visitors to identify plant species, the brochure explains the value of wetlands and their transitional nature as they slowly fill in and change from open water to forest. As you follow the trail, you will first see the remains of the open water, largely overgrown by cattails, jewelweed, and arrowhead. These provide habitats for frogs, insects, and hummingbirds. From there you move into a swampy area, with its wet soil and larger vegetation. Mushrooms, mosses, and ferns are part of the nutrient recycling process, as mushrooms feed on decaying wood, and mosses gather soil around them in which plants will later grow.

The transition between the wetland vegetation and the trees growing on the adjacent higher land provides a better wildlife habitat than either would on its own. The wetland provides a source of food, while the forest provides better shelter.

Birds inhabit these wetlands almost year-round, with the red-winged blackbirds returning in March and not leaving until late October. Ducks and Canada geese stop here on their spring and fall migrations, and the red-tailed hawk occasionally soars above the swamp in search of frogs and small mammals.

Besides the Wetlands Nature Trail, a boardwalk crosses

another marsh area, which is reached from a trail beginning between the Orientation Center and the picnic shelter.

The park is managed by the Winooski River Park District, Ethan Allen Homestead, Burlington, VT 05401; 802-63-5744.


## OAKLEDGE PARK

Between a popular park with a bicycle path and Lake Champlain, this rocky stretch of shoreline is our favorite place in Burlington to watch the sunset.

*Directions:* From US 7 (Shelburne Road) on the south side of Burlington, take Flynn Avenue west toward the lake. You'll find it north of the I-189 interchange, just past a shopping plaza. Follow it until it ends in a parking area just past Blanchard Beach. The point is on the other side of the pavilion. Or you can reach the park from the Burlington Bike Path. The cove is accessible to wheelchairs via the bike path.

A narrow strip of steep shoreline lies between the well-used park and the lake, crisscrossed by trails and studded with giant fireplaces half hidden in the low woods. Below this is a cobbled cove, protected from the wind and waves of the main lake by a narrow point of land. The point is almost solid rock, of a reddish color, its horizontal layers broken and worn away to form natural steps and benches, perfect places to sit in the late afternoon and watch the lake.

A few stunted junipers, small cedars, and a pine tree are the only vegetation of any size, their branches windblown and twisted. To the north lies the shoreline of the city and to the south a wooded point. Bluebells, fragile-looking things for such an exposed place, cling to the rocks.

On the opposite side of the cove along the shore that rises

above the rocks, is a tangle of trails opening out to frequent views of the lake. Trees here cling to very thin soil between rock outcrops where tiny sedums grow. The area is small, a ridge bounded by the lake and a ballfield, so you can't get lost, but it seems much farther away from the busy city than it actually is.

A map of the shoreline bicycle route is published by the Burlington Parks and Recreation Department, 216 Leddy Park Road, Burlington, VT 05401. Copies are available at the information kiosk on Church Street or at city parks.

## EAST WOODS

Composed of some of the richest flora of any area of its size in northern Vermont, East Woods was saved from cutting by a determined neighborhood, with help from the University of Vermont.

*Directions:* From US 7 (Shelburne Road) at the south end of Burlington, take Swift Street east. It intersects with US 7 almost directly at the I-189 interchange. The entrance to the woods is a short distance on your left; a small parking pullout is just past the sign. Just inside the woods is a signboard with guides to the nature trail.

Only here and there does the sun penetrate the interwoven branches of East Woods to reach the forest floor along which the trail leads. Evidence of human activity blends with the natural; visitors who are adept at "reading" a landscape will quickly recognize an old railbed through one section of the woods.

Fifty-six species of woody plants have been identified here. Maples and white pine trees have reached unusual sizes, but the most remarkable trees are the giant hemlocks growing along the steep slope above Potash Brook. Hemlocks thrive in sandy soil with little sunlight, exactly the conditions of East Woods, but are seldom left to grow to the size seen here. The sand on which these hemlocks stand is left from the postglacial period (about 10,000 years ago) when Lake Vermont covered most of Burlington (except, it is thought, the hill on which the University of Vermont stands today).

Near the end of the trail are several dead trees riddled with holes made by pileated woodpeckers. The trail, which takes thirty to forty-five minutes to walk, is clean and provides easy walking; only in one downhill section is the terrain steep. The University of Vermont Natural Areas Environment Program maintains the woods and invites visitors to contact its staff with any questions. You can contact them at The Bittersweet, 153 South Prospect Street, Burlington, VT 05405; or call 802-656-4055.

## LAKE CARMI NATURE TRAIL

Few places of its size in the state offer such a rich variety of wildflowers or environments as the nature trail along the shores of Lake Carmi, just south of the Canadian border. A bog with a raised roadbed through it adds still another plant habitat.

*Directions:* From Burlington, travel north on US 7 or I-89 to Swanton. There follow State 78 east to State 236; the entrance to Lake Carmi State Park is on the left. The nature trail leaves from the end of the parking lot for the beach.

The path begins by crossing a mowed area, then enters a meadow grown so high that the plants tower overhead. It skirts a marsh created by a beaver dam, passes through a lush fern bed, and comes out on the shore of the lake. It borders the lake for a short way, then enters a deep cedar forest, where the rich black

**Pileated woodpeckers hollow out characteristic rectangular holes
in trees for raising their families**

soil is often muddy and forget-me-nots carpet parts of the trail. Very little undergrowth is found in this cedar forest, where the broken branches and trunks of cedar trees have formed a web, filled in by twigs. You can readily see how cedar trees stabilize a bog and prepare the ground for larger trees to come.

After emerging into a field, the trail enters a young birch grove before coming out onto the other side of the beaver marsh. In the open areas alone, we counted twenty-six different varieties of native wildflower in bloom in August, including the rare red cardinal flower.

The road to Camping Area B cuts through the third-largest peat bog in Vermont, encompassing 140 acres. The open area adjacent to the road is a mat of sphagnum and laurel, highlighted with white puffs of cotton grass. This is a mature bog, much of it covered with tamarack and black spruce trees. There is no board-walk access, but you can see the bog vegetation from the road.

The campground here has 178 sites, 35 with lean-to shelters; several have good wheelchair access. Contact the park at RD 1, Box 1710, Enosburg Falls, VT 05450; 802-933-8383.


## TROUT RIVER FALLS

Not large enough to be considered one of Vermont's major water-falls, this is one of the few places where two falls drop almost into the same pool as a smaller stream joins the river.

*Directions:* From the intersection of State 118, 242, and 58 in Montgomery Center, go east on 242 only a few yards until you see a large yellow school building boarded over. It sits back off the right (south) side of the road, but is clearly visible as you pass. Park at the school. The trail leaves from behind the right side of the building. The falls are a five-minute hike on a good path.

The trail into the falls passes through a tumble of moss-covered boulders, two of which lie against each other to form a cave. Just past these, the Trout River falls into a natural swimming hole, where it is joined by a smaller stream cascading in from the left. Large flat rocks border the pool, and the rivers create smaller pools and rapids, where you can sit and let the water flow past. Unlike a lot of these mountain streams, this one, at least in the summer, doesn't feel like it has just melted from a glacier.

The falls and pools below it are a favorite local swimming place, the kind we hesitate to tell people about because we'd hate to see it become so crowded locals couldn't enjoy it. But the friendly owners of the Phineas Swann Bed and Breakfast there assure us that overcrowding is not a problem in Montgomery Center. It is a very attractive town, with seven covered bridges and some nice restaurants, but its remote location keeps tourist hordes from flocking there. Perhaps we should mention that the falls is a clothes-optional swimming hole for locals.

## HAZEN'S NOTCH

Hazen's is the only major notch in Vermont that is similar in shape and formation to the White Mountain notches in New Hampshire.

*Directions:* From Montgomery Center, take State 58 east through the notch, or from Lowell, take the same route west. The road is not paved and is closed in the winter.

When you read (as you will in some guidebooks) that a notch is the New Hampshire word for what is called a gap or a pass elsewhere, don't believe it. Although all of these geological phenomena may serve the same purpose to road builders – the easiest route over a range of mountains – they are distinct and different geographical features. A notch is the result of glacial scouring and is characterized by a rounded bowl on the leeward side of the glacier's path.

This happened when the flow of a glacier was restricted into a fairly narrow area in very hard bedrock, such as schist or granite. The ice was funneled by high mountains into the valley, where its abrasive forces became greater and where the valley walls and floors were more easily worn away. First, the moving ice gathered all the loose rock it had broken free as it moved, then it used these rocks to scour the surface as it continued its progress. The result of this scraping is the scoured bowl of a notch, often with walls rising in steep headwalls or cliffs.

The road rises steeply out of Montgomery Center, following Trout River for a time. Although it continues to rise, there are no long grades once the road reaches the open meadowland, where you get fine views of Jay Peak and the cliffs of the notch ahead. Although much of the way is through deep forest, views open out frequently, especially of the cliffs. Often the road-sides are hung with moss-covered rock faces, and if the weather has been wet there will be little streams dropping out of the woods. The undergrowth is lush along the road and the sun rarely penetrates.

## FAIRFIELD SWAMP WILDLIFE MANAGEMENT AREA

Created by a dam in Dead Creek, Fairfield Swamp provides a habitat for waterbirds, wildflowers, and the rare ground yew. Access into this wetland world is easy by car or, for a deeper look, by canoe.

*Directions:* From St. Albans (north of Burlington) take State 36 west to the edge of town, then go left (north) on State 104. Just after crossing I-89, take the unmarked road to your right. At 1.7 miles, take the dirt lane to the right, bearing left at the Y. This road ends in .3 mile, at the dam.

Early morning is our favorite time in places like Fairfield, when the dew hangs in droplets from each strand of the spider-webs and the birds are their most vocal. Red nightshade berries are luminous in the low backlight. Access to the swamp shore is at the dam or anywhere around the little knoll adjacent to it. The knoll itself has a unique forest floor, a ground yew whose low branches and short flat needles resemble hemlock but whose red waxy summer berries distinguish it.

The water surrounding the knoll is edged with pickerelweed and sedges and covered with both water lilies and yellow pond lilies. Wherever you approach the shore, a flurry of plops indi-cates the frogs returning to the safety of the water. Choose your footing carefully near the shore, especially as you near the beaver lodge, since you are walking on a mat of interwoven tree roots interspersed with holes.

Alongside the low dam and on the rocky islands below it, you will find tall stands of an unusual fireweed, whose single flowers don't grow in stalks as common fireweed does. Look carefully along the lower shore in summer for brilliant spikes of cardinal flower. Also blooming in midsummer are Queen Anne's lace, primrose, goldenrod, red clover, chickory, and jewelweed. The swamp is a good place to see herons and ducks.

While this side of the swamp offers both boat and land access, State 36 passes through the southern part of the Wildlife Management Area, where birdlife is plentiful.

*Of interest nearby:* On State 104, south of St. Albans, is Maple Grove Campground, a quiet, well-run spot set among tall maple trees, without a recreation center, pool, or other magnet for boisterous activity. Campers here have pool privileges at a larger

campground nearby. Maple Grove offers facilities for the handicapped, as well as easy-access sites. For reservations, contact
them at 1627 Main Street, Fairfax, VT 05454; 802-849-6439.

## METCALF POND ROCK CAVES

Giant moss-covered boulders close to the shores of Metcalf Pond
lie together, forming cavelike overhangs that once provided a
hideout for smugglers on their way from Canada.

*Directions:* Take State 36 west from St. Albans (north of Burlington) to East Fairfield. At the junction in the center of the
village, go south (right), continuing left at the fork (follow the
paved road) and left again at the T. Take the next right at the
gravel banks. In four miles, take another left, which will bring
you to the southern shore of Metcalf Pond. Park at the small
pullout by the shore and walk back up the road to a lane on the
right (north). Just before this road reaches the lakeshore summer
homes, it rises slightly. Enter the woods to your left before the
rise and walk at right angles to the road. There is little undergrowth and you will soon see a small stream bed; the rocks are
directly ahead of you, under an overgrown rock face.

The area at the foot of the ledges is only semiforested, so
you can see the rocks and explore them easily. They have fallen
from the ledge in large chunks and slabs, forming overhangs and
covered spaces between them. While these would provide scant
shelter in a hard storm, they do offer plenty of hiding places large
enough for people and goods. Local legends abound concerning
the smugglers.

From the natural history perspective, these rocks are good
examples of talus, pieces of rock fallen from cliffs or ledges.
These are in the deep shade of the forest and the mountainside

to the west, so they are covered with moss. Where they are not covered, notice the twisted wavy rock surface; in places the rocks look more like weathered and twisted cedar stumps than stone. The setting and the smuggler stories combine to make it a slightly spooky place to explore. Although the caves are on privately owned land, it is neither posted nor fenced; please show your appreciation to the owners by leaving your car at the public pullout and leaving no trace of your visit to their property.

## CAMBRIDGE STATE FOREST

Old-growth stands of white pine are rare at best, and to find them in a place with such easy access is even rarer. Naturalists estimate the age of the largest of these trees at 200 to 300 years.

*Directions:* From Jeffersonville, travel west on State 15, along the Lamoille River. Just before the road crosses the river, turn north onto Pumpkin Harbor Road. Almost immediately, take Bartlett Hill Road to the right. As that road, which turns to unpaved in a few yards, crests the first stage of the hill, turn sharp left, seemingly across someone's front yard, into Mountain View Cemetery. The forest is off the right side of the cemetery.

We can almost guarantee that you will not see the cemetery at the crest of the hill. It is behind you, on the left, and its entrance road shares the driveway of the red house. Inside, just past the little maintenance shed, you will see some of the large pines in a forest with smaller, but still impressively tall and straight, pines. These younger trees are smooth to an unusual height because they have grown in the shade of the giants that surround them and have reached straight up to get light.

The forest is on a steep hillside with very little undergrowth, and you should walk into it just to stand under and beside these venerable trees, some of which are over four feet in diameter. Before the Revolution, trees like these would have been the property of the king, to be cut as masts for the sailing ships of the British fleet. But unlike most of the fine old-growth forests of New England, these remote trees escaped both the king and later lumbering to grow to this size.

## DISCOVERY MUSEUM

Children will enjoy the exhibits and activities in this small museum, which has enough depth to keep parents interested as well.

*Directions:* Essex Junction is northeast of Burlington. The museum is located on US 2A, three blocks south of the Five-Corners in the center of Essex Junction, where US 2A, State 15, and State 117 all meet.

Outsized insect models, an aquarium, a diorama of the Grand Canyon, shell and coral collections kids can pick up and examine, and a field ecology exhibit where small visitors crawl

---

**THE FORESTS OF THE CHAMPLAIN VALLEY**
Along the borders of Lake Champlain, extending three to eight miles from the shore, is a distinctive type of northern hardwood forest. These forests are composed substantially of oak and hickory, and they occupy much of the area that was covered by Lake Vermont before it shrank to its present size. Hickory is not an especially common tree in the rest of northern New England, and these forests give an opportunity to see large numbers of them.

---

through an animal burrow and look through windows at life underground before emerging through a tree trunk—these are a few of the discoveries that await visitors to the museum. In the Lake Champlain Discovery Room children can examine marsh life and other phenomena of the lake and its shores. A small planetarium presents interpretive astronomy programs, often interweaving local history and legends of native peoples.

Changing displays supplement the more permanent ones, so it's a place you can go back to. Most of the exhibits are interactive. Workshops, field trips, preschool programs, and other regular and special events fill an active museum calendar.

The museum is open September through June on Tuesday, Friday, and Sunday 1:00 P.M. to 5:00 P.M. and on Saturday 10:00 A.M. to 5:00 P.M. July and August hours are Tuesday through Saturday 10:00 A.M. to 5:00 P.M., Sunday 1:00 to 5:00 P.M. Admission is charged. For information on the museum, contact them at 51 Park Street, Essex Junction, VT 05452; 802-878-8687.

## BROWNS RIVER GORGE AND FALLS

Beneath the historic Chittenden Mill, the Browns River shoots through a dramatic gorge, while upstream it cascades over a series of small waterfalls, where it has created potholes and other interesting rock formations.

*Directions:* Browns River crosses State 15 in the town of Jericho, six miles west of Essex Junction. Chittenden Mill is visible from the road; park at the mill to view the gorge or use the river trail.

The railings of the road bridge hide the gorge, so the casual traveler can easily cross over it without ever knowing what lies

**Kingfishers are a handsome blue and white;
the female is belted with chestnut**

beneath. Next to the road bridge, however, is a footbridge, from which pedestrians can look directly down into the rushing waters flowing between steep rock walls. In this case, the mill that was built over the gorge to take advantage of its power enhances the view, its century-old stone walls blending into the natural stone.

Curiosity will probably tempt you to cross the road and look off the downstream side of the automobile bridge, despite the fact that it is narrow and there is no space between the traffic lane and the cement rail. Don't bother; the gorge ends at the bridge and what little there might have been downstream is ruined by a broken cement dam and penstock.

Behind the mill is a trail leading upstream along the banks of Browns River to a series of cascades and small waterfalls. Side trails lead to the riverbanks, each to another stage in the continuing series of rocky falls. About a five-minute walk from the mill site is one set of falls and pools whose rocks have been worn into convoluted forms like a group of freeform sculptures. Upstream, another falls flows over layered rocks upthrust vertically so that the moving waters have worn the layers irregularly into basins and walls of different heights. Beyond is a series of rocks worn into fantastic shapes, with little basins, caves, and straight-sided potholes. Ten minutes farther on, a cascade and series of short falls includes one where the rocks have worn into a pattern that sends the water into a three-strand braid as it falls. Beside the river at this point is a rocky glen.

The entire walk is an easy half-hour excursion. The riverbanks are interspersed with small sandy beaches and pools suitable for bathing, if not deep enough for serious swimming. Picnic tables are behind the mill.

Several other short trails cover the park; follow the left-forking trail away from the river to a beaver pond formed in Clay Brook, which flows into Browns River near the picnic area.

## CANTILEVER ROCK

High on the western slopes of Mount Mansfield, halfway up a cliff face, juts a sword-shaped slice of rock about forty feet long.

*Directions:* From State 15, west of Burlington, shortly beyond the village of Jericho, take the unnumbered road east to Underhill Center. About one mile past the village, take the right turn marked "Underhill State Park." Leave your car at the entrance to the state park and continue on foot about .9 mile along a gravel road, bearing left at the intersection of the Group Camping Area road, to the Sunset Ridge trailhead. Trail maps are at the ranger station at the park entrance, but the route is well marked.

The Sunset Ridge Trail ascends steadily through the woods with only two spots that are particularly difficult when the rocks are wet and slippery. Trailside benches are placed facing scenic overlooks. At .7 mile is a junction; follow trail signs for Cantilever Rock, to the left. At about .1 mile, after passing through a narrow, rocky area, you will come to the base of a cliff. Above is Cantilever Rock, about halfway up a 100-foot cliff, sticking out at a right angle from the rock face. The area below and beyond is almost as interesting, a jumbled hillside of huge boulders that have fallen from the cliffs above. You can look up and match the pieces of talus around you to the clefts and faces of the cliff overhead, as if working on a giant three-dimensional jigsaw puzzle. And, a little disconcerting as you stand underneath, you can read the cracks and overhangs to predict which piece will be the next to fall.

Underhill State Park has eleven tent sites and six lean-to shelters. Since campers cannot drive into their sites, but must stop in the central area, the park is not suitable for trailers. Contact Underhill State Park, Underhill Center, VT 05490; 802-899-3022.

*Of interest nearby:* Mount Mansfield View Bed and Breakfast, in Underhill, has a bubble-glass solarium looking out across the entire profile of the mountain. The surrounding shrubbery was chosen to attract birds—*Rosa rugosa*, even a dead snag tree "planted" and replaced every few years to provide a leafless perch that doesn't obscure the view of the mountain. At any time we have been there, at least four varieties of birds were at the feeders or perched in view. The adjacent porch with tables and chairs is perfect for warmer weather, but from either the porch or the solarium you will have a comfortable perch for yourself and a lot of avian company. Bird guides and field glasses are handy for guests' use, as well. On Sand Hill Road, Underhill Center, VT 05490; 802-899-4793.

## PLEASANT VALLEY HERBS

A more attractive setting for herb gardens would be hard to find: Mount Mansfield provides the backdrop, and venerable maple and butternut trees frame the view.

*Directions:* From State 15, west of Burlington, shortly before the village of Jerico, take the unnumbered road east to Underhill Center. Continue through the village and look for the herb farm sign on the right. If you reach the turnoff for Underhill State Park, you have gone too far.

Patricia DeForge designed her herb beds at Pleasant Valley for looks and practicality. Set in well-manicured lawns, the beds are curved to form a half circle at the far end of the gardens, and are twenty-four inches wide. This arrangement gives them a spacious, almost formal air and also means that they are accessible to wheelchairs, which can approach directly from the level driveway. The arbor where greenhouse plants are displayed is built extra wide for easy access, and the greenhouse has a double door.

A small pond with stone edges provides habitat for moisture-loving herbs and perennials, such as mint and iris. Over 150 varieties of herbs fill the gardens, in which visitors are welcome to wander.

A small shop (with a ramp) displays herbal and herb-related products, including homemade jams and jellies from native wild berries, herbal honeys, herb blends, teas, herb vinegars, and unique herb-impressed tiles. (The local artist who creates these will also use herbs and flowers from your own garden for custom-made tiles to decorate a kitchen.)

Pleasant Valley is open Saturday and Sunday year-round, and on Friday, as well, from April through the end of September. For information on classes or programs for garden clubs, contact Pleasant Valley Herbs, Pleasant Valley Road, Underhill Center, VT 05490; 802-899-4597.

## ROCK EXPOSURES AT INTERSTATE ROAD CUTS

While it may seem odd for a book on natural wonders to suggest sightseeing on an Interstate highway, the route of I-89 passes through several cuts that give us a glimpse into the makings of the earth beneath.

*Directions:* I-89 between exit 13 in Burlington and exit 10 in Waterbury.

Naturally formed rock outcrops seldom provide us with a glimpse of the inner structure of the earth's crust. Soil, erosion, and vegetation either cover or wear way the visible evidence of the tremendous upheavals of thrust faults and the swirling effects of igneous rock creation. But I-89 required considerable cutting though solid rock, and these fresh cuts expose beautiful and dramatic formations.

Between exits 12 and 13, the road crosses a large fold in the Ordovician stratum, formed about 450 million years ago. The cuts here show a fine-grained gray marble with quartz. At exit 12 are dark brown schists occasionally striped with quartz. Exit 11 shows evidence of a thrust fault, where large blocks of bedrock have broken loose and moved sideward over other younger rock. This type of faulting puts older rock on top of rock formed later, reversing the natural order of the layers in the earth's crust. At other cuts you may notice a greenish tint to the rock, where schist is colored by the mineral chlorite.

Those with a particular interest in geology can learn more about the exposures in these and other road cuts from Bradford B. Van Diver's book, *Roadside Geology of Vermont and New Hampshire*, published by Mountain Press. It is available in bookstores statewide.

## GREEN MOUNTAIN AUDUBON NATURE CENTER

An outdoor educational center for all ages, the center maintains an active schedule of natural science events throughout the year, as well as a network of nature trails.

*Directions:* From I-89, take exit 11 and turn east on US 2. At the traffic light in Richmond, turn south toward Huntington. At just under five miles, you will pass the parking lot for the lower trails and sugar bush. Shortly beyond, Sherman Hollow Road, to the right, leads to the visitors center.

The center has displays and a library, plus a small shop. If the center is closed, look for trail maps in the small shed, below. Trails cover a wide variety of terrain and ecosystems. The

Sensory Trail provides a rope guide for the visually impaired; the White Pine Trail leads to an overlook above a beaver pond; the River Trail goes through a mature hemlock forest along the Huntington River; and the longer Tree Den Trail ventures deep into the forest to examine trees that animals are using as dens.

One of the shortest trails, which takes only about fifteen minutes, passes through beds of ten different fern species. A trail map shows the location of each and includes illustrations and descriptions to aid in distinguishing between them.

The nature center offers hikes, programs, field trips, and children's activities year-round, plus a summer Ecology Day Camp, maple sugaring (a sugar house is located on the property), and moon walks seasonally. The trails are open daily from dawn until dusk year-round. The nature center and shop are usually open weekends and most weekdays, but it is wise to call first if you want to use the library or see the indoor displays. There is no charge to use the trails or to visit the center, although they are supported through donations and the Audubon Society would welcome yours.

For membership information or a schedule of events, contact them at RD 1, Box 189, Richmond, VT 05477; 802-434-3068.

## THE BIRDS OF VERMONT MUSEUM

More than 200 species of birds, carved of wood and engraved with feather-perfect detail, fill this museum created entirely by one artist.

*Directions:* From I-89 take exit 11 and turn east on US 2. At the traffic light in Richmond, turn south toward Huntington. At just under five miles, you will pass the parking lot for the Green Mountain Nature Center. Shortly beyond, Sherman Hollow

Road, to the right, leads to the museum, which is on the right. Parking is beyond the museum, but wheelchair access to the main exhibit room and workshop is from the road.

Robert N. Spear Jr., the founder of the first Vermont chapter of the National Audubon Society, was largely responsible for the founding of the Green Mountain Nature Center, just down the road from his museum. He has carved birds since he was a teenager, perfecting his technique and taking advantage of ever more sophisticated tools, often creating his own in order to achieve a reality that is impossible even in taxidermy.

Visits to the museum begin with a short film about his work and the process of creating the birds. On the first floor of the museum is an exhibit of extinct and endangered birds of North America, including the great auk, California condor, and peregrine falcon. Here also is a replica of a long-extinct species, based on a 140 million-year-old fossil. "It's the earliest thing with feathers," Bob tells visitors.

Admission to the museum includes an invitation into the workshop where all the birds are created. Not only the birds, but the habitat settings in which they are shown—you can see how he creates each leaf from sheet aluminum and turns the eggs on a lathe. The process of engraving and coloring the feathers of each bird is perhaps the most fascinating. Since the sculptures are meant to be examined closely, not only is each feather distinct, but each barb is engraved in detail. For this he uses a woodburning pencil that he has fitted with loops of wire the same size as that used in toaster coils. By varying the heat, he can make tiny lines that do not darken the wood.

One of the many things that distinguish Bob's work from that of artists who make purely decorative birds is that he replicates the colors and shadowing, even the sheen and iridescence, of the feathers perfectly. He has developed his own technique for this, which involves heating the surface after it is painted, to blend the shades of the feathers. Those visitors interested in craft technique will find the workshop as fascinating as the birds themselves.

The birds fill the main exhibition room, each nesting pair in its own Lucite case, inside which Bob has created the bird's native environment. Occasionally he can use real plant material, like the tree trunks for the woodpeckers, but more often he must simulate the flora as well. It is hard to tell the difference. The attitude of each bird is not only lifelike, but typical of that particular species. A Lincoln's sparrow stands lightly on a bog mat of sphagnum moss, watching a fly poised on the brink of a pitcher plant leaf. The oven bird's unique nest is in a root under trillium and trout lily. Chimney sweeps are (where else?) in the corner of a chimney. The ruffed grouse is shown with a brood of nine amid their broken shells and straying onto the forest floor, where they blend in so well you have to look very carefully to find them all.

The displays are set at average adult eye level and step stools are provided for those with lower eye levels. To date, Bob Spear has completed 151 species of Vermont birds, including all the songbirds. It is not a process that can be hurried, and one feels when meeting Bob that "hurry" is not in his vocabulary. The magnitude of this project is so great, and the patience required so infinite, that it would seem that he must spend every waking hour at his workbench.

The museum is open May 1 to October 31, from 10:00 A.M. until 4:00 P.M. daily except Tuesday, and at other times by appointment. Admission is charged. Contact the Birds of Vermont Museum, RFD 1, Box 187, Sherman Hollow Road, Richmond, VT 05477; 802-434-2167.

# 2

# Northeast Kingdom

## BIG FALLS

The only large falls on a major river in Vermont that has not yet been destroyed by a dam, Big Falls is also the habitat of several rare wildflowers.

*Directions:* From North Troy, just south of the Canadian border, take State 105 east one mile to a dirt road to the right (south). The road is marked "Troy Falls Road," but the sign is located out of sight of the intersection. Go about 1.5 miles to a pullout on the right, where there is parking for the falls.

One of the state's largest waterfalls, its name describes it aptly, if tersely. The Missisquoi River, about sixty feet wide above the falls, drops through a rocky ravine in three separate channels, cascading about twenty-five feet into a bubbling froth. Immediately below the falls is a gorge with steep rock walls about sixty feet high. Downstream from the gorge is a deep pool plus sandy beaches, a popular local swimming place.

Plant life around the falls includes a rare *Erigeron*, a variety of goldenrod found elsewhere in Vermont only in difficult-to-reach places, and *Aster tradescanti*, also rare. The banks are forested with pine, hemlock, and a few hardwoods.

In case you wonder why the Missisquoi River, which you may have met at its delta as it flows into Lake Champlain, is flowing due north here, it flows into the province of Quebec, then turns west and south, back into Vermont and west to the lake.

## HALL'S CREEK MARSH

Hall's Creek hardly seems large enough to create such a wetland as it moves slowly toward Lake Memphremagog, but the low meadow-bordered marsh through which it flows is rich in wildlife and best explored by canoe.

*Directions:* Located on the shore of Lake Memphremagog in the town of Beebe Plain, north of Newport at the Canadian border. Take an unnumbered road due north from Newport, following

---

### DON'T BLAME THE GOLDENROD

Allergy sufferers have long blamed their miseries on the bright yellow goldenrod that colors the fields in August. But the real culprit is the ragweed, whose pollen is light and easily windborne. The pollen of goldenrod is sticky and cannot fly anywhere. But while the ragweed has inconspicuous blossoms that would never be noticed in the fields and roadsides, goldenrod, which blooms at the same time and usually in the same places, is colorful and easily noticed. Actually, those with allergies should be thankful to the goldenrod for alerting them to the presence of ragweed pollen in the air.

---

the shore of the lake until the road comes to a T at a cow pasture. (Don't be surprised if, as you top the rise, the cows appear to be grazing in the road directly ahead of you.) Go left until the road makes a sharp right turn at the lake. Just past that turn, the road passes between the lakeshore and Hall's Creek Marsh on a narrow causeway. Park at the pullout for the boat access to the lake. There is no sign.

Hall's Creek Marsh is connected to the lake through a culvert under the road and lies, like Lake Memphemagog, along the border. In its cattail-lined streams you will see the work of beavers, hear the songs of a number of birds, glide under willow

**Both parents raise otter pups, teaching them to swim, hunt, and play**

trees and among yellow pond lilies. The marsh continues along the border for quite a distance, between gently sloping pastures and forests. In addition to its bird population, the marsh is a habitat for otter and muskrat. You can put a canoe into the small stream beside the road, which leads into the marsh.

## PROSPECT HILL OBSERVATORY

An observation tower, the work of the local Historical Society, offers sweeping views across farmland and forest to the Green Mountains and the Northeast Kingdom.

*Directions:* In Brownington, left off State 58 in Orleans, follow signs to the Old Stone House Museum. The entrance to the observatory is next to the white church at the intersection in the center of the village. It is clearly marked with a sign.

The observatory is a two-story wooden tower at the top of Prospect Hill, from which you can see a 360-degree view out over the surrounding landscape to hills and mountains in every direction. In the foreground are farmlands in the valleys. For permission to camp overnight in the field at the crest of the hill (you can't take a car up there after dark without permission), call the Orleans County Historical Society, which built the observatory, at 802-754-2022 or 754-6657.

## WILLOUGHBY FALLS

One of the few places in Vermont where you can see steelhead trout in their annual spring migration, the falls of the Willoughby River in Orleans provides a good vantage point.

*Directions:*  From the center of Orleans, take Church Street to E Street and turn left. You will see the falls from there.

Each spring, usually in April, the steelhead rainbow trout migrate from Lake Memphremagog, where they have spent the previous year, upstream to Lake Willoughby for spawning. To do so, they must negotiate the small waterfall in Orleans, where the river drops about four feet in a series of cascades and pools created by several layers of ledge in the river. The configuration of the shore is such that spectators can stand within a few feet of the fish, jumping and at rest in the pools.

The small riverbank area is protected by the Nature Conservancy. For information on the migration, call the Vermont Department of Fish and Wildlife at 802-241-3700.

## LAKE WILLOUGHBY

Certainly the most dramatic of Vermont's many lakes, Willoughby is set in a steep cleft between two mountains whose cliffs rise steeply from its shores.

*Directions:*  To reach Lake Willoughby from I-91, north of St. Johnsbury, take exit 25 and go northeast on State 16. To reach the southern end, take US 5 north from Lyndonville, then 5A in West Burke. US 5A borders the narrow eastern shoreline of the lake.

One of the deepest lakes in New England, Willoughby was cut by a glacier from an already existing valley. During the advance of the glacier, the valley was scoured into the deep-sided shape we see today. As the glacier moved, it picked up bits of rock it had torn loose from local bedrock on its way. As it began to melt, it dropped these pieces, along with finer gravels and debris it had picked up in its travels. In this area, the debris formed dikes, blocking both ends of the lake and allowing the water to accumulate to an unusual depth.

As the glacier melted, it left behind plants that thrived in its Arctic-like environment, but did not thrive in the warmer climate that followed. Most of these plants died, but in a few places they found just the cool atmosphere they needed. One of these places is on the cliffs that overhang the lake on Mount Pisgah. These alpine plants are relatively safe, since they grow from the most inaccessible crevices of the cliffs, far from the nearest trail. Sweet broom, mountain saxifrage, and a number of other very rare plants grow there.

The cliffs are also a nesting place for peregrine falcons, so trails may be closed during nesting season in the spring. The best time to see falcons and hawks is during migration in the fall, during late September and early October. Later, from late October into November, large buteos move through the area. Especially when the wind is from the west, sightings can be quite numerous from the top of Mount Pisgah, just north of the summit from an overlook to the north.

In addition to the raptors soaring at high elevations, birders in the forest may see warblers, boreal chickadees, cedar waxwings, ruffed grouse, and goldfinches.

The dramatic cliffs of Mount Pisgah and Mount Hor can be reached by hiking trails. Parking for the trail along the face of Mount Pisgah is south of the lake at the intersection of an unpaved road on the west side of the road. Parking for the Hawkes Trail to Mount Hor is up that dirt road on the right, where the trail begins. Several other trails give access to views out over the lake and adjoining mountains. Since some of these are not too well marked, you should as a minimum have with you the trail map published by the state Department of Forests, Parks and Recreation (180 Portland Street, St. Johnsbury, VT 05819;

802-748-4890). *Day Hiker's Guide to Vermont* has even more detailed information on the trails, but not as large a map. This small book is published by the Green Mountain Club and is available in bookstores throughout the state.

## NORTHEAST KINGDOM NATURE TRAIL

Hills and lakes ground down and scoured out by glaciers were dominated by the railroad and lumbering industries from the mid-1800s until well into this century. Brighton State Park captures the essence of a boreal forest and the impact of man on it.

*Directions:* Follow State 105 south of the town of Island Pond along the eastern shore of Island Pond. The park is located along the shore of Spectacle Pond and is accessible from State 105 and from a road that passes between these two ponds. A trail map is available at the ranger station.

The town of Island Pond gained in importance from the time of its founding in 1853 as the rail hub of the Grand Trunk Railroad. The railroad opened up the northern parts of both Vermont and New Hampshire, encouraging not only the establishment of farms but the exploitation of the vast tracts of virgin timber that the area contained. This is the heart of the Northeast Kingdom, a vast, sparsely populated area serviced by few roads and encompassing the bulk of the state's northeast highlands. The forests here are composed of evergreens, coniferous trees that include balsam fir, red spruce, red pine, and white cedar. An important aspect of the forest trails here is the insight they give into the logging industry, which employed over 500 loggers here in the early years of this century. The stumps of enormous white pines are still visible, as are "walking" trees, trees with roots extending above the ground as a result of the rotting away of the stumps from which they originally grew. The self-guided nature trail is half a mile long, and a longer one-mile hiking trail is also accessible. The terrain is easy.

The centerpiece of Brighton State Park is Spectacle Pond, a kettle hole created by a large chunk of ice left trapped in glacial debris by the retreating ice cap. The ice created a basin as it melted, only eight to ten feet at its deepest and with no significant stream feeding it. It has no major stream inlet. Around this depression, forests of giant white pine trees grew until the land was nearly cleared by logging in the early years of the twentieth century. On the park's half-mile nature trail, one of these giants remains; look for it at the point labeled "9."

In the 1850s, when the railroad opened this area, there were still signs of Indian encampments and trails. The point of land now known as Indian Point showed evidence of having been a gathering place, since it had been cleared except for some of the tall pines that shaded its landward end.

The predominantly evergreen forest that covers the vast area of northeastern Vermont is known as a "boreal forest" which, in simpler words, means northern woods. Balsam fir, red pine, red spruce, and white pine are common and, in the moist, boggy areas, tamarack and white cedar. The major hardwoods are maple and birch, with aspen an early regrowth in cut areas. These trees are the most common throughout the park, as well as the surrounding woodlands.

Beyond the nature trail, which begins at the small museum close to the campers' beach, hiking trails wind throughout the park, often ending at old logging roads. The trail to Indian Point is about a mile long. Wildflowers common to the area are bunch-berry, checkerberry (wintergreen), and the fragile goldthread and twinflower; several varieties of ferns grow in the woods, and the forest floor is carpeted with club moss. In the morning, it sounds as though birds fill every tree, and at night you may hear the cry of a loon visiting the pond. It is not unusual to see moose in the park.

The campground offers sixty-three tent sites and twenty-one lean-to shelters, each surrounded by woods, with a few bordering the pond. All of the lean-to shelter sites are wheelchair accessible, and one is reserved exclusively for that use. Evening nature programs offer films, talks, and nature games. Other than a bathhouse at the campers' beach, there are few frills, making this a favorite camping place for those who love the quiet, the wilderness, and the chance to live close to nature. Those who enjoy fishing will appreciate the region's clear lakes and fast-moving streams.

---

### IS THAT A MOOSE?

The largest animal in the northern woods, the American Moose (*Alces alces*) is a magnificent mammal to behold. Standing over nine feet tall at the shoulder, its mass can be overwhelming at first sight. Although moose are usually seen alone, small family groups of a cow and a calf may be encountered. They prefer low wet areas and congregate around streams, ponds, and marshes, where their favorite foods—willow, water lilies, and other aquatic plants—flourish. If a road passes through a low marshy area, particularly in the northern part of the state, you may well find moose grazing at the margin of the dry and wet lands. While they may seem timid and gentle, *don't be misled*. These giants can become a raging terror in no time if they feel threatened, and what may seem nonthreatening to you may seem like a panzer attack to them. The best rule is to give them plenty of space, move very slowly, be very quiet, and let your zoom lens or binoculars bring you closer. When driving at night in the areas where they can be expected, be extra vigilant. They don't know fear because they have no natural predator. Your car is just a strange animal to them, and they will walk out right in front of you, so be prepared to brake whenever one is within sight of the road.

---

Trail maps are available at the ranger station at the park entrance. The park is open mid-May through Columbus Day weekend. Contact Brighton State Park, Island Pond, VT 05846; in summer 802-723-4360, winter 802-479-4280.

## MOOSE BOG

Deep in the thick coniferous forest of the Wenlock Wildlife Management Area, Moose Bog is a fine example of a floating bog surrounding a pond.

*Directions:* From Island Pond, north of St. Johnsbury, take State 105 east for almost eight miles. Shortly after you cross the railroad track, there is a sign for the Wenlock Area on the right as you come to the end of the spruce forest. Just a mile past the sign, turn right onto South America Pond Road. If you are arriving from the east, look for a sign on the right, almost hidden in the woods, and take the left turn directly opposite the sign, onto South America Pond Road. Drive .2 mile and look for a trail entrance on the right blocked with a pile of stones. Park well off the road to allow trucks to pass.

The trail into the bog is easy to follow, since it is an old logging road. Finding the trail down to the bog is a little harder. As you walk, notice the small clearings, actually just widened places in the trail. You should reach the fourth one after twelve to fifteen minutes of walking (that includes time to pick a few blackberries, put on more insect repellent, and inspect the lichens by the trailside). Look there for a trail to your left, which descends shortly to the bog.

You will soon see the pond through the trees; the trail makes a sharp left turn then emerges along the shore. You will immediately notice that the shore is a false one, actually a floating mat of interwoven roots, fallen twigs, and other vegetation. It jiggles like pudding underfoot. Be careful where you step, not only because bog water is cold when you step through into it, but because your footprints can damage the fragile plants and the base they grow on.

Bog laurel, which blooms in late June, extends from the tree line to the edge of the water, interspersed by other plants, including leatherleaf, bog rosemary, Labrador tea, and, of course, sphagnum moss, which forms the deep carpet from which the other plants grow. Everywhere, their shiny leaves conspicuous enough, but their round red blossoms even easier to spot, pitcher plants are waiting for a lunch of insects. Unlike many bogs with high boardwalks, you are at ground level here, so you can inspect the pools in the leaves closely to see the leftover parts the plants could not digest.

Most visitors come to Moose Bog for the birding, which is extraordinary, with ten species on Vermont's rare, threatened, and endangered list. Walking along the path you may encounter a spruce grouse, which is a breeding species here, as are the black-backed woodpecker and Canada jay. The boreal chickadee, yellow-bellied flycatcher, gray jay, Tennessee warbler, and white-winged crossbill are also seen here.

While the variety of vegetation in bogs is somewhat limited by the acidity of the environment, each bog is different and each bears exploring. But this is one of the few where the mat is still growing out from the edges and the pond is still open water in the middle. It is among our favorite bogs, for this reason and for its utter remoteness.

## MOOSE VIEWING

Moose have returned in quantity to the Northeast Kingdom, partly because timbering operations have left cutover areas,

## BOREAL FOREST

Coniferous (cone-bearing) trees dominate the typical northern forest that covers almost the entire northeastern highlands area of the Northeast Kingdom and the northern sections of the Vermont Piedmont, to its immediate south and west. The area extends roughly north from St. Johnsbury and from the Connecticut River to a point a bit west of Island Pond. This type of forest girdles the northern part of the globe up to the Arctic tundra line, across Canada and northern Russia, where it is called the taiga. The soils are gravelly and very poor, most of the minerals leaching out with groundwater and thereby limiting the plant species that will grow. Evergreens, mostly spruce and fir, dominate from an altitude of about 3,000 feet to as high as 4,000 feet, where the trees become gnarled and stunted, no more than two or three feet tall. This high-altitude region is called a krummholz zone, the German word for crooked tree.

which provide the best habitat for wildlife. Deep forests don't allow in enough light for the lower-story growth that provides browse. Moose are most active late in the afternoon, in the evening, and in the early morning. As you travel the roads in this part of Vermont, look for low marshy places beside the road. Very often these will have areas of deeply churned dark mud, a sure sign that moose have been feeding there.

Marshy places and the shores of shallow ponds (you can tell shallow ponds by the amount of vegetation in the water, such as cattails, pond lilies, or pickerelweed) are the best places to watch. It is not unusual to find several cars stopped along the roadside in such a place; they may be watching a moose or simply waiting for one to appear. Although local people see them in the road all the time (and you probably will, as well), moose watching is a favorite evening activity. Always brake for moose, however, for they have no fear of your car and will step right in front of you. It's a contest you will lose.

**Moose browse for tender twigs and water plants
along the edges of swamps, ponds, and lakes**

If you see moose, be sure to stay in your car, unless they are far away. They are huge and, although not normally dangerous to humans, they are still wild animals and may misunderstand your friendly interest, especially if they have calves with them.

The Wenlock Wildlife Management Area (see directions for Moose Bog) has several moose-viewing places along State 105. Another prime area is along State 114 north of Island Pond and along the Canadian border between Norton and Canaan.

*Of interest in the area:* If you would like to spend a few days enjoying nature up close or fishing in one of Vermont's most secluded lakes, go to Quimby Country. This old-fashioned fishing camp is like none other. Its cottages face the shore of Forest Lake, which, except for Quimby's, is surrounded by boreal forest. This is a land of moose, muskrat, and loons, virtually unchanged since Quimby's first opened its lodge over 100 years ago. The same families have been coming here for generations, but the first-time visitor is welcomed like an old friend. Contact Quimby Country, Forest Lake, Averill, VT 05901; 802-822-5533.

## BRUNSWICK SPRINGS

Rich in legend, the six springs flowing from the steep riverbank have a turbulent history.

*Directions:* From Bloomfield, east of Island Pond, go south on State 102 for about two miles. Park at the small town hall on the left side of the road and walk down the dirt lane beyond it until you reach a pond on your right. The springs are down the steps to the left, toward the river.

The rusted pipes, leaf-clogged pool, and cement steps are a forlorn remnant of what was once a popular healing spa. The springs were known to the Abenaki for their curative powers, and, according to local lore, the Abenaki placed a curse on anyone who tried to exploit the springs for profit. There are as many versions of the story as there are people to tell it, but the consensus is that every time someone tried to build a resort here, it burned. Some stories tell of the Abenaki bringing a wounded British soldier here to cure his injured arm, and all agree that the springs became increasingly popular throughout the nineteenth century. People seeking cures stayed as boarders with local families until a small hotel was built. It burned to the ground, although not on the eve of its opening, as some accounts relate. Another followed, also destroyed by fire. A grander resort was planned and nearing completion when it, too, burned. Its foundations remain today, half covered in riverside vegetation.

What of the six springs, each reputed to be of different mineral content? The smell in the air suggests that sulfur predominates in at least one of them. Despite the abundant testimonials to their effectiveness, none of the curative claims have been supported by scientific evidence. It's a slightly eerie place, with its ruined hotel overhanging the steep bank above the river sixty feet below.

## MAIDSTONE STATE PARK

The most remote of Vermont's state parks, Maidstone is one of the best places to spot loons, which nest and raise their young on the lake.

***Directions:*** From Bloomfield, which is east of Island Pond, take State 102 south along the Connecticut River. At about five miles, turn west onto a dirt road; it is marked by a sign to the park. The ranger station at the entrance is another five miles. Maidstone Lake was formed much like Lake Willoughby, when the glacier carved more deeply into an existing valley. There is still evidence of the glacier: gravel, sand, and large boulders dropped by the melting ice. To see some of these erratics, as glacially deposited boulders are known, take the three-quarter-mile "Moose Trail" through the forest between camping areas A and B. The trail leaves the road in area A near campsite 22 and goes through the woods and up to the crest of a hill, where many of the boulders are found.

Another, shorter, trail along the shore of the lake is a good place to look for loons, as well as a beaver lodge. It begins near campsite 35 in area B and goes through the woods to the lake. Along with loons, Maidstone Lake is a good place to see black-backed woodpeckers, evening grosbeaks, wood warblers, and pine grosbeaks. In the spring, look for pink lady's slippers along the woodland portions of the trails and for the rarer white lady's slipper.

Campsites in the park are well spaced with a good variety of secluded and open spots. Several lean-to sites are suitable for handicapped campers; one is reserved especially for them. Contact Maidstone State Park, RFD 1, Box 455, Guildhall, VT 05905; 802-676-3930.

## VICTORY BASIN (VICTORY BOG)

A prime example of the boreal forests of the northeast highlands of Vermont, Victory Basin Wildlife Management Area also contains bogs, marshes, and hardwood forest.

*Directions:* From St. Johnsbury, follow US 2 east to the village of North Concord. Turn left (north) on an unnumbered road marked "Victory" and travel about four miles (don't take the left-hand road marked "Victory Hill"). The wildlife management area boundary is marked with a sign.

Moose River parallels the road through the basin, which is ringed by knolls and hills of from 1,200 to 1,500 feet in elevation. Trails and old woods roads provide access from the "main" road. Most of the area consists of spruce and fir, the trees that define the true boreal forest, with bogs and marshes in the lower depressions. This whole area lies on a granitic base, sandy and acidic. What nutrients are added by decaying plant matter are soon leached away in the sand. This inhospitable floral habitat is further limited by the cold climate and short growing season, a classic subarctic environment.

In the bog areas, typical bog vegetation, such as pitcher plants, Labrador tea, and leatherleaf, thrive. In the forests of red spruce and balsam fir, there are few understory plants, and the forest-floor vegetation exhibits only those plants that flourish with little light. Partridgeberry, goldthread, bunchberry, trailing arbutus, ferns, and club mosses predominate, and lady's slippers in both pink and the rarer white bloom in the spring.

Birdlife varies throughout the year, with juncos and white-throated sparrows nesting here in numbers, along with waterfowl such as mallard, black, and wood ducks. The basin supports a great-blue-heron rookery and transient populations of birds on their way to nesting sites in Canada. Even in winter there are birds here, as the finches, including grosbeaks, come south in search of food. Moose are seen here frequently, and about 1,000 acres of the managed area is deer yard.

About 1.4 miles past the Wildlife Management sign is a pulloff on the left side of the road. Almost hidden in the woods is a stone marker with a plaque dedicated to Fred Mold, through whose efforts Victory Bog became a protected area. "He cared about the least of his brethren," the memorial reads, "and took the time even to feed the birds." For this wild and peaceful place, where the spires of fir trees stand above the forest and the sounds

**The white-throated sparrow sings** *Poor Sam Peabody, Peabody, Peabody*

of birds and insects break the stillness, we can pause to thank its protector.

*Of interest in the area:* Among the loveliest of all the wonders of nature in Vermont are the sunsets, but unless you are looking westward across a lake, the hills often make sunsets difficult to see in all their glory. The Wildflower Inn in Lyndonville seems to have been built to solve this problem. At the crest of a hill, its gardens and porches overlook a pastoral valley and what seems like more sky than any one place is entitled to. The inn welcomes families, with special programs for children, including nature walks, as well as abundant nearby hiking trails for all ages. The gardens alone are worth a visit. The inn is on Darling Hill Road, Lyndonville, VT 05851; 802-626-8310 or 800-627-8310.

## FAIRBANKS MUSEUM

Discoveries for all ages fill this remarkable institution, created and maintained by people who find wonder in the smallest snowflake and the mysteries of the universe.

*Directions:* US 2 and 5 and I-91 and I-93 all converge in St. Johnsbury. From I-91, take exit 21 to US 2 (Western Avenue), turning left on Main Street. From US 5, go west onto US 2 and right onto Main Street. The museum is at Main and Prospect Streets.

It would be a shame to pass the Fairbanks Museum off as a Victorian period piece full of mounted birds, fossils, and snowflake photographs. It is far more than that, although the building is indeed a classic example of the best Victorian museums, with a high vaulted ceiling and wooden display cases elegantly crafted of golden oak and cherry. The building itself is worth seeing, with its arches, towers, and elegant portico of limestone and red sandstone, built by a local benefactor, to display collections of everything that fascinated the inquiring nineteenth-century mind, from a polar bear to Masai shields.

The older exhibits, although they are not presented with modern interpretive techniques, are informative and interesting. Newer ones explore the ecosystems of Vermont, help visitors identify local wildflowers, and examine weather trends and phenomena (the coldest weather in this century was in 1943, when the temperature dropped to −43 degrees; only three times has it reached 101, all in the early decades of the century).

The Fairbanks Museum is the home of the state-of-the-art weather station that has for more than 100 years recorded, and with the advent of radio, predicted the most unpredictable of all events: the Vermont weather. It was Fred Mold who initiated the radio program based on the museum's Northern New England Weather Center. The programs, now presented by Mark Breen and Steve Meletsky, still retain the early style Fred Mold brought to them, with astronomical and farming information woven into the scientific data. Those New Englanders who plan their day with "Eye on the Sky" each morning will enjoy seeing the studio and weather station on the museum's ground floor.

Fred Mold, whose memorial we visited in Victory Basin, was a man of great vision and a director of the Fairbanks Museum. It was also under his leadership that the museum's planetarium was conceived and constructed. Unlike most planetarium facilities, this one does not depend on recorded or computerized narration. Live voice programs give the museum staff the chance to create varied and lively programs based on current or past astronomical events as well as a chance to incorporate other subjects, such as bird migrations, legends, or local history, into the programs.

The museum is open from 10:00 A.M. to 4:00 P.M. Monday through Saturday and 1:00 to 5:00 P.M. on Sunday. During the summer, those hours are usually extended. Planetarium shows are at 1:30 P.M. on Saturday and Sunday year-round and daily during July and August. Admission is charged to both the museum exhibits and the planetarium shows. The museum address is Main and Prospect Streets, St. Johnsbury, VT 05819; 802-748-2372.

## STODDARD SWAMP

In a cedar bog, where the sphagnum and the small woody plants it supports have created a footing strong enough for trees, their roots and fallen wood in turn further strengthened the mat.

*Directions:* From US 2 in West Danville, west of St. Johnsbury, go south on an unnumbered road just west of the intersection with State 15. At about 4.9 miles a woods road enters from the right, with a barred gate. Park there and walk down the hill on the other side of the road; you will see the cedar trees ahead of you. If you miss the woods road and come to a T at Ewell Pond, just turn around and go back .7 mile.

There are no trails into the Stoddard swamp, but walking down the lightly forested hillside is quite easy. Once in the cedar bog, you will have to pick your way carefully, testing each step to avoid wet feet; waterproof boots will be handy here, but are not essential.

The atmosphere here is like no other bog we have visited. Full-sized cedars grow thickly, their roots woven together and reinforced by fallen branches and trunks of dead trees into an uneven floor, the spaces filled in by rich green sphagnum. Moss has grown over the roots and fallen trunks as well. Through the area a small brook runs, its bed a soft gray sand. A tree that has fallen over the brook becomes a bridge for animal traffic and is soon overgrown with a layer of moss, a trail worn along its center, with ferns bordering its edges. In places the mat of fallen trunks and branches has covered the brook entirely.

The floor, created where a rotting stump or trunk has broken down into dark, humus-rich, but acid soil, is carpeted with wood sorrel. Iridescent blue dragonflies dart among the ferns and low evergreens. This is a mature bog, carbon-dated to over 11,000 years old, and you don't have to go very far into the bog to realize and enjoy its unique environment. If you plan to photograph here, take high-speed film, since the upper canopy is almost as tightly woven as the mat beneath. Sunlight penetrates in vivid splashes, but only occasionally.

## GROTON STATE FOREST

Over 25,000 acres in the southern Northeast Kingdom, the park encompasses ponds, several mountains, one of the state's largest and most diverse bogs, and vast stands of second-growth forest. Within its boundaries are four campgrounds and a group camping area.

*Directions:* State 232 runs through the center of Groton State Forest, between State 302 in Groton on the south and US 2 in Marshfield on the north. The entire area lies close to and southwest of St. Johnsbury. The following places are all signposted from State 232. Trails are marked with blazes of blue paint and, occasionally, orange diamonds.

An outstanding feature of the park is its publications, which include lists of birds sighted and plants known to grow here, each coded for season and location. Another booklet describes the history of the area, including the railroad and logging days. It is well illustrated with old photographs. For information about the State Forest, contact it at Marshfield, VT 05658; 802-584-3820.

### Nature Center and Trail

Between Big Deer Campground and Boulder Beach, at the northern end of Lake Groton, the Nature Center sits at the beginning of an interpretive nature trail. The center itself contains a museum with displays of local plants, birds, habitats, and geology. The latter is explained in a series of diagrams showing where the Nature Center site was throughout the geologic history of the area. Its displays are a good way to understand the many changes, from the bottom of a shallow sea and the rising of magma to erosion by glaciers. The floor is paved with examples of Vermont's rich heritage of stone – granite, limestone, slate, and others, each labeled.

The nature trail begins behind the center – pick up a booklet with notes on the stations along the trail. The trail begins with further explanation of the ground underfoot, its formation, and the first and most primitive plants, the lichens. It explores the varied terrain, from the hardwood forest recovering from extensive logging to the northern spruce and fir forest. The emphasis is on learning to read the landscape and understand its evolution from its shape and the plants and animals that inhabit it.

## Peacham Bog

The trail to the bog begins at the far end of the parking lot where the Nature Center is located. It is a fairly long hike (allow at least two hours) but a particularly interesting one. The terrain is mixed, from low woodland and a raspberry thicket to the rocky slope of a fairly steep hill. The bog sits at an elevation, one of only two known "raised" bogs in the state. It is surrounded by a coniferous forest of tall trees with almost no understory. At one point, the environment seems almost like a rain forest, with moisture-loving plants and mosses.

The bog itself does not have a boardwalk, so waterproof boots will be very helpful, especially in wet seasons. Because the bog is fragile and because you are far from traveled areas, you should stay on the trail.

Along with the plants frequently seen in bogs—pitcher plants, leatherleaf, Labrador tea, and bog rosemary—you may find several orchis varieties, including the purple fringed and white fringed. Birdlife is rich in the boreal species, including the rare black-backed three-toed woodpecker, several flycatchers, boreal chickadees, and at least two warblers. Several varieties of thrush, kinglets, the solitary vireo, crossbills, and a number of different sparrows inhabit the woods through which the trail passes.

## Owl's Head

Geology and a fine view of the mountains and ponds are the main interests in the short (ten-minute) climb up the CCC-built steps from the parking area to the top of Owl's Head. Or you can climb from the base via a trail from the road between New Discovery Campground and Osmore Pond. Birders will often be rewarded with hawks circling overhead.

The conspicuous ledges of light-colored granite at the top of the mountain (and others in the park) were formed over 300 million years ago, but until the retreat of the last glaciers, about 12,000 years ago, they were not exposed. As the glaciers passed, they scraped away the softer rocks covering the granite. More of this same bedrock lies beneath this whole part of Vermont, but is covered by layers of glacial deposits, topsoil, and vegetation. If you visited the Nature Center first, you have seen a graphic explanation of this geological history.

What is particularly interesting about the granite at the top of Owl's Head is the appearance of dikes, bold stripes that cross its surface and each other. After the granite had solidified from its molten state, cracks appeared in it. These were filled in one of two ways, either by magma (molten rock) that was forced into the space and cooled there or by the crystallization of minerals from water that settled in the cracks. Thus the stripes are actually solid sheets that extend well below the surface and far beyond the edges of the exposed rock.

The coarser of these two dikes is composed of pegmatite, which contains the same minerals as granite—feldspar, quartz, and mica—but often in large crystals instead of finely blended. The lighter-colored dike is aplite, containing a higher concentration of quartz, which accounts for the white color, but in very fine grains. At the point where two of the dikes cross, their relative ages become apparent: the pegmatite is younger, since it crosses through the aplite. (Not all stripes in granite are formed this way; some are inclusions, which are slices of older rocks, such as schists, that were captured and held by the cooling magma when the granite was formed.)

## MARSHFIELD FALLS

Possibly the longest continuous cascade in the state, the falls continue for about 100 feet.

*Directions:* From US 2 in the center of Marshfield, about halfway between Barre and St. Johnsbury, take Depot Street south. It turns off just west of the Rainbow Sweet Cafe and Bakery (where you should not resist the temptation to stop for Danish, lunch, or a wedge of Engandiner Nusstorte). Go either way at the fork—each shortly crosses a bridge over the cascade.

Unlike most Vermont waterfalls, Marshfield Falls flows over a granite ledge. As the stream drops, it spreads out over the granite face into dozens of tiny falls and cascades. Each bridge offers a different view. Looking upward from the top bridge, you see the stream appear out of the woods, already whitewater. From the lower bridge, you can look down onto the point where it changes from a cascade to a series of straighter drops into pools.

In the spring it changes from a little brook cascading over the rocks to a mass of foam and spray. Unlike so many falls that you cannot get to safely in the spring, this one has two viewing platforms. The brook joins the Winooski River just below the falls.


## CHICKERING BOG

Perhaps the most fascinating thing about bogs is their variety; no two are the same. Chickering is open and unshaded, with a raised boardwalk for access to its interior.

*Directions:*  From US 2 in Plainfield, north of Barre, go north on State 214 to North Montpelier. Turn right onto State 14 and go about a mile to a crossroad, marked by a sign for the Calais Elementary School. Go left, and at 1.5 miles pass George Road. At .1 mile farther on, you will see a driveway on the left and a very tiny marker on a green post labeled "TNC." Park along the road, *not* in the driveway, which leads to a private home. Walk up the driveway to a woods road, which leads straight ahead when the driveway makes a sharp right turn. Follow the woods road downhill, bearing left after a low, marshy area. A trail marked with the TNC initials goes off to the right and is well blazed from that point. A trail map is posted as the trail approaches the bog. It is about a twenty-five-minute walk from the road to the edge of the bog.

Fifteen acres of quaking bog mat surround a tiny pond in what is known as a "gramminoid fen" bog. When you look at it from a distance, it could be a flat grassy meadow with a few small trees beginning to grow. But a closer look at the vegetation

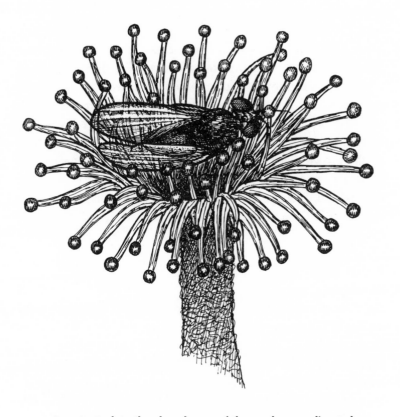

**Insects stuck to the gluey leaves of the sundew are digested
by the plant to add nitrogen to its diet**

proves its origins. Along with pitcher plants, you may see the
rare livid sedge, rose pogonia, sundews, bladderworts, sweet
gale, grass pinks, blue flag, and several rare orchid species.

A fen differs from a true bog in that it has some flow of
water through it, although often very slight. So instead of having
no oxygen and an acid water, it may have some oxygen and
perhaps a neutral or even base water. This accounts for some of
the differences in plant species—sedges instead of sphagnum

predominate and the presence of tamarack instead of cedar—but many of the species are the same, since the two types of bog share many of the same environmental problems for plants. Absorption of nutrients from the water below is difficult in both.

In Chickering Bog, an occasional tamarack grows ten to twelve feet tall, but they are widely spaced in the rare places they can find a foothold. Especially interesting in the early spring when the orchids are in bloom, the bog has its charms at any time of the growing season; in late July, the dark red flowers and shiny leaves of the pitcher plants stand out against the filmy background of sedge. The bog is managed by the Nature Conservancy, 27 State Street, Montpelier, VT 05602; 802-229-4425.

## BARR HILL NATURE PRESERVE

High above Caspian Lake, this nature preserve combines a wide variety of plant life with interesting views.

*Directions:* From State 15, east of Hardwick and west of St. Johnsbury, take State 16 north 4.8 miles to Greensboro Bend. Turn left on an unnumbered road to Greensboro. Take a left at the T and an almost immediate right. Bear right at the fork, following the sign for Vermont Daylilies, and continue uphill past the lily fields to the electric fence. The road passes through a farm to an electric fence that crosses the road. Disconnect the fence and drive through, immediately reconnecting the fence before proceeding. Follow the sometimes rough track to a parking area near the top of the hill. A trail map is available at the beginning of the trail or from Barr Hill Nature Preserve, The Nature Conservancy, 27 State Street, Montpelier, VT 05602.

This 256-acre preserve offers two trails, a short one of .3 mile and a longer version .6 mile in length, both of which are on easy grades. Seventeen stations highlight the unusual features of this important property. A site of farms as little as a century ago, the forested hilltop is a good example of second-growth forest and natural reclamation of the land. The well-written trail guide explains the origin of the quartzite deposits visible at station A as well as the origins of Caspian Lake, visible from station C. One station has a view from which you can see a spectrum of mountains spreading from Signal Mountain and Groton State Park on the south to Mount Mansfield and Madona Peak in the west. The hilltop includes fine examples of the effects of the last glacial period, 15,000 years ago, including scouring and gouging and erratic boulders carried here from miles away by a glacier and deposited when the glacier melted away. There are several good places to picnic, but camping is not permitted. All materials brought in must be carried out—and don't forget to replace the electric fence on the way out.

## RUNAWAY POND

There isn't much left to see of this pond; it moved north in 1810. Man tinkered with nature but, read on . . .

*Directions:* On State 16 between Glover and Greensboro Bend, about halfway between Clark's Pond (a mile to the north) and Horse Pond (to the south), both on the west side of the road and each identifiable by a public boat access sign. Watch for a small sign, also on the west side, by a low marshy area.

First, to set the stage: the Barton River rises in Clark's (sometimes called Tildy's) Pond and flows north through Glover and Orleans to Newport and into Lake Memphremagog. The Lamoille River, before our story begins, rose in Long Pond, just a mile south of Clark's, flowing first south, then west to Lake

Champlain, as it still does. (If you are following this on a map, don't be confused by the other Long Pond farther to the west, which doesn't figure in this story.)

The mills on the Barton River needed more water to operate efficiently, and Long Pond seemed a likely source, since it was a mile and a half long and half a mile wide and sat at an elevation higher than the river. On June 10, 1810, a group of about sixty local men gathered at the northern end of Long Pond with shovels to dig a small channel, allowing some of the water to flow north into Clark's Pond and thence into the Barton River.

They dug and the water began to flow. What they didn't know was that Long Pond was supported on the north by nothing more than a ridge of fine sand lined by clay, which had formed a layer like the rim of a bowl sealing the end of the lake. When that clay rim broke, the weight of the lake full of water quickly shattered the remaining edges of the clay and washed away the sand barrier that had supported it. Long Pond emptied—some reports say in a mere fifteen minutes, others that it took several hours—in a wall of water between fifty and sixty feet high. Instead of powering the mills, it washed them away, along with homes, farms, and everything else in its path. The Barton River now begins in the marshy area that was once the bottom of Long Pond, and the Lamoille rises a mile farther south, in Horse Pond.

## A GARDEN TOUR

The region from Marshfield to Glover is extraordinarily rich in perennial gardens and plant nurseries. All in Hardiness Zone 3, which has a short growing season even in good years, these patches of brilliant bloom give hope to anyone with a northern garden. Although other natural wonders lie between these gardens, we have grouped them into a linear tour, only referring to other attractions that are described separately. Since several of these are in remote locations, we will not repeat the directions from St. Johnsbury each time, but will direct you from each garden to the next. With a map in hand (which you will always need in Vermont, since road signs are so scarce), you should have no trouble finding individual gardens if you choose not to visit them all.

## Vermont Flower Farm (Marshfield)

*Directions:* From US 2 in Marshfield, west of St. Johnsbury, take State 232 southeast toward Groton State Park, making a left turn onto Peacham Pond Road just north of the park boundary. The gardens are immediately visible on the right.

Before you see the field where the stock plants bloom, you will see the driveway lined with rows and rows of pots from which sprout a profusion of Asiatic and trumpet lilies in every color you can imagine (except blue). Before the lilies, come the spring bulbs; perennial border flowers and lilies are at their peak mid-July through August.

The farm is open 10:00 A.M. to 4:00 P.M., July 9 through late August, daily except Monday. The address is Peacham Pond Road, RR 1, Marshfield, VT 05658; 802-426-3505.

## Amanda's Greenhouse (Cabot)

*Directions:* From US 2 in Marshfield, follow State 215 north toward Cabot. Amanda's is on the left before you reach Lower Cabot.

The sloping fields of this wide valley are solid green in the summer, except for those surrounding Amanda's. More annuals, especially rare and unusual ones, grow here than at the other gardens in this tour. Look for showy nolana and calliopsis.

Another specialty is the selection of late-blooming lilacs, especially attractive to people with frost-prone locations.

The gardens are a joy to wander in all summer long and are open May through September every day from 9:00 A.M. to 6:00 P.M. You can contact Amanda's Greenhouse at 74 Cabot Road, Marshfield, VT 05658; 802-426-3783. As you drive north after leaving, watch for their delphinium fields beyond the next house.

### Henry Menard's Gardens and Bird Carvings (Cabot)

*Directions:* From State 215 in the center of Cabot, take Danville Hill Road east, up the hill, and watch for a sign on your right.

This is certainly the most unusual garden you will see in northern Vermont. To your left as you ascend the driveway appears a forest of topiaried conifers, like a bonsai collection grown up. Set around a landscaped pond with an island, rocks and perennial plants, like hosta, astilbe, and cascading potentilla, are mixed with the evergreens. Contorted weeping spruce, tamarack, native spruce, even white pine have been trained to unusual shapes that include spirals, urns, cones, and poodle cuts.

Once a nursery and perennial farm, the gardens are now a hobby for Henry Menard, who is a bird carver as well. His lifelike birds are ornamental, perched on stands in realistic attitudes. The birds are for sale, beginning at about $20, and he will happily show them to you, but you might wish to call first and be sure he is at home, at 802-563-2501.

### Perennial Pleasures (East Hardwick)

*Directions:* From Cabot, continue north on State 215 to State 15, then west to State 16. East Hardwick sits barely off State 16, to the west. Follow the main street down the steep hill, over the bridge, and straight up the other side (don't turn right when the main road does). Perennial Pleasures is on your left.

If you are interested in restoring an old garden, plan to spend some time here. The owners, who have worked for some time in garden restoration, found it difficult to locate the right plants for different periods and resolved to open a nursery dedicated to historic perennials and herbs. The gardens are obviously the work of a well-trained professional. Enclosed by walls is a tea garden where they serve a real English cream tea from 2:30 to 5:00 every day except Monday, from Memorial Day to Labor Day.

The catalog ($2.00) contains a gold mine of plant and herb information and advice on restoring old gardens, as well as a source by mail for those who cannot come to Vermont to get plants. For a copy or to reserve a room in their authentic English bed and breakfast, write to 2 Brick House Road, East Hardwick, VT 05836; 802-472-5512.

## Vermont Daylilies (Greensboro)

*Directions:* Continue north on State 16 to Greensboro Bend. Turn left on an unnumbered road to Greensboro (there is a Vermont Daylilies sign at the intersection). Take a left at the T and an almost immediate right. Bear right at the fork, where you will see another sign, and continue uphill until you see the lily fields.

"Daylilies," according to David and Andrea Perham, "are the perfect perennial, ideal both for dedicated gardeners and for those who love beautiful flowers but have neither the time nor inclination to putter over their planting." In mid- or late summer, their hillside is covered with splashes of yellow (thirteen different shades we counted), orange, pink, red, peach, orchid, and white—a total of 400 cultivars. If you are going to Barr Hill Nature Preserve (see above) and also plan to buy lilies, we suggest you choose the varieties first, then go to Barr Hill while the plants are being dug and labeled, stopping for them on your return.

The nursery is open every day except Monday, from 10:00 A.M. to 6:00 P.M. between July 1 and Labor Day. To see the lilies at the peak of bloom, plan to visit from the last week in July through August. The Perhams also operate a small, homey bed and breakfast at the farm. To reserve or request a plant list, write to Vermont Daylilies, Barr Hill Road, Greensboro, VT 05841; 802-533-7155.

### Stone's Throw Gardens (East Craftsbury)

*Directions:* Return to the village of Greensboro and make a sharp right at the fork at the base of the hill and continue past Highland Lodge (a good place to stop for lunch on the porch overlooking Caspian Lake) and into the village of East Crafts-bury. When the road makes a turn to the left in the village, go straight ahead. Stone's Throw Gardens are on the left, a short distance up that road.

The display gardens are set into a landscape of stone walls and foundations, along fences, and up a gently sloping hillside in a steady progression of color. Nursery plants in large pots are arranged in rows and beds amid those in the ground to create even thicker masses of bloom, especially lush in mid-July, when the roses are still blooming, the summer perennials are at their best, and the lilies have begun to bloom. Roses, primrose, delphinium, foxglove, salvia, clematis, bluebell, black-eyed Susan, phlox, lamb's ears, veronica, bee balm, achillea, and monkshood are backed by the Green Mountains in the distance. Stone's Throw is a testimonial to what a lot of hard work can accomplish even in a Zone 3 garden.

The gardens are open Wednesday through Sunday from 10:00 A.M. to 5:00 P.M. from mid-May through late August. For information: at Route 1, Box 1320, Craftsbury, VT 05826; 802-586-2805.

## Labour of Love Nursery (Glover)

*Directions:* From Stone's Throw Gardens, return to the center of Craftsbury, but take the left turn just before reaching the main road. This unsurfaced, and poor, road leads through South Albany and West Glover to State 16 in Glover. The gardens are in the center of the village of Glover.

If the side yard of this 1880s home on Glover's main street were a lawn, it wouldn't be so big that the kids would complain about mowing it. But every inch is awash with the color of over 300 different varieties of perennial flowers, from borders of tall delphinium that form the backdrop for an English cottage garden at one edge, to a shady space on the north side of the house planted in lush shades of green contrasting with delicate white and pastel blooms. Don't expect formal walkways, pergolas, and sundials. Do expect a lot of flowers in full bloom from April until hard frost.

The gardens are open Thursday, Friday, and Saturday, until dark, other times by appointment. Write to Labour of Love at HCR63 #44, Barton, VT 05822; 802-525-6695.

# 3

# Central Mountains

## CADY'S FALLS NURSERY

An old river-valley farm has become a nursery with show gardens of perennial plants.

*Directions:* From State 15 in Morrisville, just west of the junction with State 100 south, take Needle Eye Road south a short distance to a T. Turn left down the hill, then right immediately after crossing the bridge. The farm is at the end of the public road.

The landscaped gardens at Cady's Falls Nursery show plants in a variety of settings. The first displays large clumps of showy perennials, such as aconitum, lilies, bee balm, poppies, and mallow, in semiformal beds among small trees and ornamental shrubs. The garden drops slightly to a lily pool and a low rill, planted with moisture-loving varieties. Rising on the other side in gentle terraces is a garden of conifers, some trimmed to standards. Among these stylized evergreens are weeping birches, ferns, sedums, and a few flowering and foliage plants arranged with a distinct Japanese flair.

The gardens are open in May and June from 10:00 A.M. to 6:00 P.M. daily except Monday. July through mid-August they are closed both Sunday and Monday. For an appointment to visit after mid-August, contact Cady's Falls Nursery, RD 3, Box 2100, Morrisville, VT 05661; 802-888-5559.

## DOGS HEAD FALLS

Although the vertical drop of these falls is not dramatic, it is the only falls on a wide river in Vermont that has not been spoiled by a dam or by other human evidence.

*Directions:* From the center of Johnson, leave State 15 to the south on the only road that crosses the Lamoille River. Immediately after the bridge, turn left and follow the road through the lumber millyard to where it begins to rise. Park in the little road to the left.

You will hear the falls from the road and can easily find the trail leading through the narrow fringe of low growth and out onto the rocks that border the falls and the wide pool at its foot. There is almost no sign of human activity as you stand on the bank here, except the people fishing from the rocks. The river narrows from seventy-five feet at water level to only ten, feeding through two stone dikes and falling in a six-foot falls into the pool. It then flows through a central channel, where it cascades from pool to pool.

Be sure to notice the bleached sticks, limbs, and entire tree trunks caught in the rocks all the way to the top of the side ledges. These show how high the spring floodwaters rise. Because the falls are open and the river wide, there is no mossy, moist ecosystem here. But the rocks bordering the falls are covered with lichens, which give them a mottled effect.

## BREWSTER GORGE

A small clear stream flows over and through a jumble of enormous boulders caught in a cleft of rock walls.

*Directions:* From Jeffersonville take State 108 south toward Smugglers Notch. Immediately after the Grist Mill, easily recognized by its large mill wheel, turn left (east) on Canyon Road. Just before the disused covered bridge, turn right into the parking area. Park here unless you have a high-wheeled vehicle, in which case you can continue to the second parking area.

The trail begins at the far end of the car park and continues through the woods beside Brewster River for about 100 yards. At that point the trail crosses the river, which is only a few inches deep in the summer (those with hiking boots can probably make it across with dry feet). About fifty yards beyond, the gorge rises directly ahead in an open area. Unlike most falls, this one has sufficient space and daylight for good photography, although that makes it a less hospitable habitat for mosses and moisture-loving plants.

At the top, the river cascades into two deep potholes, one almost ten feet wide. The gorge is as deep as fifty feet in places. The pool at the bottom is a popular local swimming hole. Because of the necessity of fording the river, this trip should not be attempted during the spring runoff, when the river will be much deeper and faster.

## SMUGGLERS NOTCH

A more dramatic mountain pass would be hard to find, as the road winds among giant boulders under cliffs as high as 1,000 feet.

*Directions:* The notch lies north of Montpelier, on State 108 between Stowe and Jeffersonville.

Having made the distinction between a notch and a gap in the description of Hazen's Notch (see page 25), we should explain that Smugglers Notch does not fit the glacial scour definition of a notch. Not every mountain defile called a notch is a glacial scour; as names, the terms notch, pass, and gap are used interchangeably. The origin of Smugglers Notch is thought to be glacial, however.

The most accepted theory, backed by geological evidence, is that as the last glacial age was coming to an end, the glacier at the eastern (Stowe) side of the notch melted first, while the Lake Champlain glacier to the west remained. As it melted, the bulk of the glacier prevented water from flowing westward, so it carved an outlet through the Green Mountains, forming the gap. The shape of the land at the top, which is V-shaped—typical of river cuts—instead of U-shaped, as glacially worn passages tend to be, supports the theory. So does the winding nature of the notch; a glacier passes through in a straighter path. Deposits of glacial sediment add further evidence.

In the millennia since the notch's formation, pieces of the cliff have fallen into the bottom of the V, forming the tumble of talus chunks that make the higher parts so dramatic. The cliffs to the west (the Mount Mansfield side) are especially uneven, making them more susceptible to the weathering forces that cause chunks of talus to break off. One of these, known as King Rock, has a sign explaining that it fell from the cliffs overhead in 1910 and that its weight is estimated at 6,000 tons (this is not a misprint).

Rock caves can be found in the spaces between the talus chunks, the largest of which is Smuggler's Cave. While there is not a lot of space for cars at the top, some parking is available and it is worth exploring the area on foot. A clear spring, called Big Spring, issues from the rocks south of the height of land.

The cliffs of the notch, in addition to their unique appearance and geology, are home to at least a dozen rare plant species, some not found anywhere else in Vermont. These alpine plant species, remnants of the ice age, thrive in the cool, moist environment of the high-altitude cliffs. Several, such as the ferns, also thrive on the calcium provided by the rocks. The unique combination of mineral-rich groundwater, limited topsoil, and cold creates what is known as a Cold Calcareous Cliff Community, which supports its own flora.

Although plants that get their nourishment from trapped insects are usually associated with bogs in Vermont, another carnivorous species lives on the cliffs of Smugglers Notch. The butterwort's tough sticky leaves capture and digest insects. Other rare plants found among the cliffs are alpine sweet-broom, purple mountain saxifrage, marble sandwort, felwort, alpine woodsia, and pale painted cap. Hikers and rock climbers need to be especially conscious of the fragile nature of the rare plants here and avoid stepping on them or dislodging them from the cliffs.

In the summer, a ranger-naturalist is stationed at the summit and can answer questions about the rare plants, as well as provide illustrated leaflets describing them. For further information, contact the Department of Forests, Parks and Recreation, Waterbury, VT 05676; 802-828-3375. Vehicles pulling trailers are not allowed in the notch—you'll see why when you drive through it — and the road is closed in the winter because of the difficulty of plowing it.

## MOUNT MANSFIELD ALPINE TUNDRA

Only two places in Vermont support plant populations common to the Arctic tundra; the largest of these is the 250 acres of alpine habitat on Mount Mansfield's long ridge.

*Directions:* Mount Mansfield lies north of Montpelier, alongside State 108 between Stowe and Jeffersonville. The summit can be reached by car on a toll road or by gondola, both from State 108 north of Stowe. It can also be reached by any of several hiking trails, three of which have trailheads on State 108. The Long Trail crosses this road in Smugglers Notch, leading to the northern portion of the summit, known as the chin. Those who have taken the Sunset Ridge Trail from Underhill State Park on the western side of the mountain to Cantilever Rock can continue up that trail to the chin. Do not attempt to climb Mount Mansfield on any of these trails without a copy of *The Guide Book of the Long Trail* published by the Green Mountain Club.

Of all these routes, the shortest and easiest way to get to the tundra area is by the gondola, which has its terminus on the long flat ridge between the nose and the chin. (Various points along the summit are identified according to the profile it faintly resembles.) The sedge tundra is at the chin. To see all the various environments of the summit, however, you will need to walk the entire ridge from the chin to the toll road (the nose).

Although 1,000 miles south of the vast tundra vegetation zone of the Arctic, the summit of Mount Mansfield provides much the same environment: shallow soil, low temperature, high wind, heavy rainfall, and a short growing season. This tundra community, and the smaller one on Camel's Hump, formed as the glaciers melted. These tundras once covered a far greater area of recently thawed land, but as the glacier left, its cooling influence diminished and the climate grew too warm for these cold-loving plants. Only on the exposed summits did conditions remain hospitable and the tundras survive.

Looking much like a grassy meadow, the tundra's predominant vegetation is Bigelow's sedge, a tall plant with broader leaves than most sedges. In the lower areas where moisture gathers, small peat bogs have formed, with sphagnum moss and plants common to bogs at lower altitude, such as leatherleaf, hare's-tail grass, and bog laurel, as well as crowberry, an Arctic plant found this far south only at high altitudes. In patches of gravelly soil you will see the mountain sandwort, common to many New England summits. Still different plants are found south of the toll road; these include willows and diapensia. Although the unique flora is the primary interest here, you may see a few birds, including the raven and, during fall migrations, hawks.

While it is tempting to get as close as possible to these rarely encountered plants, remember that although they are resistant to severe weather, they are very sensitive to foot traffic, especially that of boots. Stay on the trail or on exposed rock surfaces and do not wear hobbed boots, which are particularly damaging to both trails and vegetation. A ranger-naturalist is on duty during the summer and can answer questions about the plants. The summit of Mount Mansfield is owned by the University of Vermont, but a number of agencies cooperate in preserving its rare and fragile environment.

For more information on alpine flora, as well as illustrations to help you identify plants in the field, *The AMC Field Guide to Mountain Flowers of New England* is invaluable. Small enough to fit easily in a pocket, it includes both alpine plants and those found at lower trail elevations.

## BINGHAM FALLS

Falling and tumbling through a narrow gorge, this falls could
well be the finest in the state.

*Directions:*  From Stowe take State 108, also known as Mountain
Road. Continue on for about half a mile past the entrance to the
toll road, to a parking space on the left side of the road. Across
the street a trail leads to the gorge and falls.

The path through the woods follows a moderately easy
grade until it nears the falls. When you reach the gorge, follow it
upstream (left) where the walls of the gorge are only three feet
high. This section has outstanding examples of potholes, holes in
the bedrock of the stream caused when pieces of rock were
caught in a swirling eddy of water and washed around in a circle
until they wore a pot-shaped hole in the rock and themselves into
flecks of sand. The number, size, and variety of potholes here,

**A pair of great blue herons on their nest platform of sticks**

some as big as two feet in diameter, make this one of the best sites in the state to observe this phenomenon.

Downstream, the gorge becomes narrow, deeply cut, and very precipitous, the water falling and cascading over rocks and through crevices, ending in a cascading fall about twenty feet tall before pouring over an abrupt drop of the same height into a thirty-five-foot-wide pool at the bottom. Above the water rise sheer craggy walls of schist and gneiss, with mosses clinging to the bits of soil that have managed to get a foothold. The whole site is set in a deep forest with a clear understory.

Although a spectacular sight, these falls are among the more dangerous we have visited. The trails beside the stream traverse extremely steep hillsides. The many people who have visited here have worn away the delicate plant life that holds the soil in place, causing a lot of hillside erosion. The trails tend to be moist and therefore slippery. Exposed roots often provide the safest means of climbing and descending to the lower sections. While the upper section can be viewed with little difficulty, exploration of the lower sections should be done with extreme care. For your own safety, and to protect the fragile banks from further erosion, it is important to use a route well away from the falls when you go downstream to see the lower falls. The trails are not as steep there and you will have better footing.

*Of interest in the area:* Set on a hillside overlooking the Green Mountains, the Trapp Family Lodge is a legend for reasons far beyond the romantic story of its founders. From its beginnings the lodge has maintained its close-to-nature atmosphere, protecting the 2,000 acres of northern forestland around it and making it accessible with hiking and 40 miles of cross-country ski trails. Before eco-tourism became a watchword, they were finding ways to make the resort environment-conscious, and they continue to lead in that endeavor. The Trapp Family Lodge is on

Trapp Hill Road, Stowe, VT 05672; 802-253-8511 or 800-826-7000.


## MOLLY BOG

Several stages of bog evolution are present here, and although there is no boardwalk or other access to the mat itself, the sloping edges provide a good viewpoint.

*Directions:* From Stowe, drive north on State 100 to the Stowe-Morristown line marker, just past Randolph and Tinker Roads, on the west side of the road. Park in the pullout beside the road and walk west along the north side of the field to the trail at the far side.

The trail into Molly Bog continues due west from the end of the field, straight through the saw-yard, where the ground is spongy with accumulated sawdust. After a ten- or fifteen-minute walk from State 100, the trail enters a small clearing. To the left, slightly below the trail, you will see the round shallow bowl of the kettle bog. A few steps bring you to its unsteady wet sphagnum floor.

Kettle-hole ponds were created by segments of frozen glacier that were covered and insulated by debris and melted after the rest of the glacier had gone. Molly Bog has formed over such a pond and is one of Vermont's older bogs. All the classic features are visible: open water in the center, a quaking mat of sedges, an encircling mat of shrubs, including laurel, leatherleaf, and other heathers, surrounded by a forest of scrubby coniferous trees. A good sampling of bog vegetation is present here, and the birdlife is richer than in most bogs. It is owned by the University of Vermont and protected by The Nature Conservancy. For more information, call 802-656-4055.

### CREATION OF A BOG

Bogs have long been places of mystery and uneasiness for humans, full of strange occurrences and unknown dangers. The term boogeyman comes from the sense of dread they cause. Actually, bogs are merely bowls, gouged out of the rock by glaciers, that have held a body of water over thousands of years and over which plant life has grown until it created a mat covering most, if not all, of the surface area. Because water is trapped in a bog and doesn't flow through, little oxygen is present and organic matter is slow to decay. Bogs have their own species of plants that thrive in this environment of acidic moisture. Primary among these is sphagnum moss, which forms thick beds over the water, creating the base on which other plants begin to grow. As mosses and other plants begin the cycle of growth and decay on top of the water, they in turn form soils that support new life. As you walk on top of some bogs the ground will shake beneath your feet. These are the so-called quaking bogs that are actually built out over an underlying body of water. Some of Vermont's bogs are as old as 14,000 years.

## MOSS GLEN FALLS

One of Vermont's longest and most dramatic waterfalls, Moss Glen is also one of the easiest to reach.

*Directions:* From Stowe, go north on State 100 for three miles to Randolph Road. Go right, then take the first right, at .3 mile, onto Moss Glen Falls Road. A short distance from the turn, there is a parking pullout on the left, with a hand-lettered sign to the falls.

The trail leads through a low, moist meadow, which can be quite muddy in the spring or after a rain. The path is bordered by tall, lush growth, including fragrant ferns. In about 100 yards, the trail enters the woods and almost immediately thereafter a side path leads to the river, where the foundation of an old mill is visible in the banking. A little farther along the main trail, a second path leads to the river, this time along its banks to a place where, at low water, you can cross to a sandbar and continue into the deepest part of the gorge directly at the foot of the lower (and longest) falls. If the water level allows, you should go to the base of the falls to see the moss-covered rock face and cavelike formations.

While there, look up to the left and note how deeply under-cut the wall is above the gorge and how the mossy forest-floor matting of shallow tree roots drapes over the edge. You can also see trees with their roots against the rock face and their tops in the riverbed, where they fell when their roots were no longer able to cling to the rocks. Great sharp sections of schist lie in a tumble where they have broken off the sides of the gorge. It is an excellent example of how such gorges are formed and grow from a weak place or a crack in the upthrust rock.

It is also a graphic picture to have in mind when you return to the main trail, climb to the top of the gorge, and are tempted to hold onto a tree and take a step closer to the edge for a better view. At the top, notice that you are at the edge of a narrow ridge; you can see one place where the edges of the rock strata are exposed. The view from here is of the whole falls, with a shorter straight drop at the top above the lower spreading cas-cade. For the sake of the fragile, steep terrain, which suffers erosion when it is subjected to heavy use, use the gentler path-ways to the left instead of the straight-up route closer to the edge.

## GREEN MOUNTAIN CLUB

The headquarters for hikers and climbers, the Green Mountain Club is responsible for the building and maintenance of the remarkable trail system the length of the state.

*Directions:* The Club Building is located on the west side of State 100 in Waterbury Center, just north of Waterbury.

Anyone considering hiking or climbing in the Green Mountains should start here. The selection of maps, guides, and books about Vermont, its trails, and its wildlife is the best we have found anywhere. If you need the map of an obscure trail, they have it, along with a full range of nature and field guides, books about wilderness travel, even booklets on canoeing Vermont's various rivers. They offer detailed and reliable information; their staff members are hikers and climbers who've probably been where you're going.

A small museum displays photos of the Long Trail's construction, a relief map of the Mount Mansfield area with trails marked, the gear suitable for day hiking, and a quirky collection of mountain artifacts: the evolution of the packboard, things found on the trails, even a paint-can belt worn by trail volunteers replacing blazes.

The club is a sponsor of the Ranger-Naturalist Program, which provides trained personnel at the most fragile and sensitive natural sites to educate visitors on proper use of the area. The center is open from 9:00 A.M. to 5:00 P.M. every day. For membership information or to order books by mail, contact the Green Mountain Club, Inc., RR 1, Box 650, Waterbury Center, VT 05677; 802-244-7037.

## CAMEL'S HUMP ALPINE TUNDRA

There is no easy way to see this remnant of early Vermont's chilly past—you have to climb the mountain.

*Directions:* Go south on State 100 and US 2 in Waterbury, remaining on State 100 when they separate south of town. In less than .1 mile, take the road to the right, signposted for the elementary school. Almost immediately, as that road makes a sharp left turn, go straight ahead on the dirt road. At just under five miles, take the road to the left and follow it along Ripley Brook, bearing left over the bridge at the fork, until the road ends at Couching Lion Farm and the trailhead parking. Be sure to review the trail map at the hiker sign-in near the beginning of the trail so you will know which of the many trails to follow to the summit. The climbing distance is four miles to the summit via the Forestry and Dean Trails, then along the Long Trail. You can shorten the descent to 3.4 miles by returning on the Forestry trail alone. The climb is a moderate one; the descent on the upper part of Forestry Trail is steep.

Although the alpine tundra at the summit is the goal of the trip, you will pass at least one other "natural wonder" on your way. The hardwood forest through which you hike at the lower altitudes is in a near-virgin state, with maple, beech, and birch predominating and occasional red oak and hickory. Higher up, as the maples thin out, you will see the transition to balsam fir trees, common to boreal forests. You will also see red spruce, although balsam is more common. As the trail climbs higher, these trees grow increasingly scragglier, until you reach the krummholz zone (see the box "Boreal Forest" in section 2) of stunted and twisted trees closest to the summit.

At the summit is a small area of Arctic vegetation, able to survive only because of the high elevation. Although the tundra here is only ten acres, it contains significant plants, such as Bigelow's sedge and mountain sandwort. As you hike, remember that rare and fragile sedges look very much like common grass and treat each plant you see as though it were endangered (which is a pretty safe bet up here). Step only on the trail or on rock out-crops and be careful that your boots do not kick smaller stones loose. Wherever the alpine vegetation is removed or dies, and wherever a rock is moved, bare soil remains. The high winds on these peaks will quickly strip away that little bit of soil, leaving no foothold for a future plant and baring the roots of neighboring ones. For this reason, we suggest that you not wear hiking boots with lugged soles, which are much harder on trails and vegetation.

The mountain is owned by the State of Vermont and, for more information, you can contact the Department of Forests, Parks and Recreation, Waterbury, VT 05677; 802-241-3678. Before you climb, we suggest that you get a copy of the *Guide Book of the Long Trail*, available at the Green Mountain Club headquarters in North Waterbury or at bookstores.

## STEVENSON BROOK TRAIL
## AT LITTLE RIVER STATE PARK

Farmed as late as seventy years ago, the land along the brook is now deep in forest, and the nature trail shows how to read the landscape for evidence of its past uses.

*Directions:* Take US 2 west from Waterbury and turn north onto the road marked "Little River State Park." To reach the trail, go left at the park headquarters (where you can get an interpretive trail guide) to the nature trail parking pullout, on the right.

Along with the human past, the geological history of the area is visible in the landscape. The terraces created by the stream as it flowed through the valley at different levels are easy to see, rising like giant steps. The interpretive booklet explains their formation as well as the shaping of the rounded stones in the brook bed. Some of these are not water-worn, but are concretions, small circles of calcium deposits formed around some object (possibly the stem of a plant) in the clay of the riverbank. The bank itself is composed of silt deposited by the last glacier.

In a casual walk through these woods, some of the evidence of the farms that were once here will be obvious—a cellar hole beside the trail and a stone wall. Others may not be so easy to recognize, such as the apple trees now shaded out by taller trees, or the "cabbage pines," sure signs that this was once cleared land. Cabbage or wolf pines are those large white pines with numerous branching side trunks, which can only take that shape in open land. Those in the forest grow straight and without large branches, so the presence of cabbage pines shows that they must have grown when the land was cleared of trees.

Little River State Park has an active schedule of nature programs, boat rides on the reservoir, wild-edibles walks, canoe trips, and evening slide shows on local wildlife. The "Meet the Trees Tour" explores the rich variety of the forests and ends at the park's prize tree—a huge hemlock estimated to be over 300 years old. If that tour is not offered during your stay, you can see the hemlock to the left of the road just before campsite 44.

Campsites here are wooded and well spaced, many of them overlooking the waters of the reservoir. The dam that created this impoundment, however, ruined one of the state's finest waterfalls; you can see the ledges over which the river once poured, on the far side of the river below the dam as you enter or leave the park.

For information on Little River State Park, write or call RD 1, Box 1150, Waterbury, VT 05676; 802-244-7103 in the summer or 802-479-4280 in the winter.

*Of interest in the area:* If you've dreamed of a cabin deep in the woods, where you can spend your evening watching the forest animals from your own front porch, Grunberg Haus Bed and Breakfast in Duxbury might have just what you're looking for. High on the hillside above their Austrian chalet, the owners have built a few—very few—cottages. No automobile sounds will disturb your slumber here, since you can't even drive your own car to your door. Be prepared to carry your luggage up the short, steep trail and bring a flashlight if you decide to enjoy the con-viviality of the main chalet in the evening; there are no street lights to guide you home. Grunberg Haus, RR 2, Box 1595, Waterbury, VT 05676; 802-244-7726 or 800-800-7760.

## MORETOWN GORGE

This is not one of the great gorges of Vermont. There are many bigger or wider or grander, but this one is human scale, and you can walk downstream for a water-level view of it.

*Directions:* From State 100, south of Waterbury, take State 100B just north of the Moretown town line. In less than half a mile the road takes a turn to the right, passing over a bridge. Immediately over the bridge is a parking space just off the left side of the road.

Follow the overgrown road beside the river for a short dis-tance, then follow a trail to the river. Following the trail farther takes you to a steeper section that descends over the foundations of an old mill and down onto the riverbed. On the way, stop to sample the wild blackberries if they are ripe. Purple flowering raspberries abound here, too. The gorge, fifteen to thirty feet wide with vertical rock walls twenty-five feet high, offers a blend of boulders and water-worn stone as well as a swimming hole.

After looking at the gorge from this perspective, continue on along 100B to the village, where a left turn leads to a bridge over the river with a nice view back along the gorge.

## BARRE GRANITE QUARRIES

The sheer size of the Rock of Ages quarry is impressive enough; then you realize that the space was once occupied by solid granite.

*Directions:* Traveling west through Barre on US 302, go south (right) onto South Main Street (at the Veterans Memorial—

carved of Barre granite, of course) then left onto Quarry Street, which leads to Graniteville and the Rock of Ages Quarry. You will see signs to the visitors center. Signs also point the way from US 302 east of Barre.

The first granite quarry in America began business here just after the War of 1812. Barre granite's uniform texture, even color, and medium-sized grain, caused by slow, even cooling from its molten state, is especially good for monument carving. The surface takes a high polish and can be carved into fine detail, and it soon became the preferred stone for a nation busily at work constructing and embellishing monumental buildings.

The visitors center offers a tour on weekdays, but the quarry and its displays are open weekends, as well. Behind the center, whose facade is faced with examples of granite in a number of colors, is the quarry. The figures—1,200 feet long, 600 wide, and 350 deep—simply don't prepare you for how big the hole in the earth is. The marks where blocks have been removed create the illusion that the walls of the pit are built of giant blocks of stone.

Monument work is exacting, and only about 15 percent of the quarried stone is perfect enough to market. The other 85 percent accounts for the mountains that surround Websterville and Graniteville. On closer look, these are huge mounds of waste stone, or grout—the stone that couldn't be used.

For those who are interested in the story of the granite industry, the visitors center has a small book entitled *Carved in Stone* that discusses the quarries from their beginning and is illustrated with old photographs.

## GRANVILLE GULF

The gulf is a deep V-shape between a long ridge to the east and a line of mountains to the west.

***Directions:*** The gulf is easy to find on State 100, northeast of Rutland, along the main road between Warren and Granville. A sign marks the entrance to the state-protected land.

Cool and shaded in the summer, when the sun rarely penetrates the deep coniferous forest, the gulf is even more dramatic in winter, when the ledges and rocks along its sides are coated in the blue ice of springs and small waterfalls frozen in motion. A stand of virgin (or at least very old) hemlock and red spruce is on the eastern slope.

Moss Glen Falls I (to distinguish it from the other one of the same name north of Stowe) drops over a rock face into a pool right beside the road. If you were designing a waterfall and were talented, this is the way you'd plan it. Its wide, rounded rock face spreads the water into a balanced flow, white against the dark mossy wall. A narrow gorge has worn through the rock, almost perfectly centered above the falls and oriented so that the late afternoon sun filtering through the forest above creates a golden glow visible through the dark rock on either side. And below, a dark shallow pool spreads from the splashing water to your feet.

The only thing to ruin the scene is the necessity of posting a large warning sign detailing the fatalities and injuries of individuals who have attempted to climb the falls.

Parking for the falls is just to its north. A cascade, also quite pretty, falls near the parking lot, and one could easily assume this to be Moss Glen Falls. Continue south on the path beside the road for a few more yards and you will come to the foot of the "real" falls.

## WHITE RIVER TRAVELWAY

A microhabitat lies along the White River and its tributaries as they flow from the eastern slopes of the Green Mountains.

***Directions:*** State 125, 73, and 107 parallel the east-flowing tributaries; State 100 follows the White River itself. The area lies midway between Rutland and Montpelier.

Rivers provide wildlife with a common route for their migrations and for the travels required in their everyday life; even plants migrate along rivers. The thousands of relationships that develop along the White River and its tributary streams have created a microhabitat, a complete ecosystem different from that of the surrounding land.

Weather, water, soil, rocks, and plants all combine in creating this microhabitat, and each creature that is drawn to the river's environment in turn brings others. The spring melt-off brings enormous quantities of stone, soil, and sand downriver, undercutting banks en route and creating new sandbars, shoals, and even islands as the waters slow. These attract creatures such as the wood turtle, which travels from its forest home to lay eggs in the sand, and the spotted sandpiper, whose preferred food source is the insect life of these bars and shallow shoals. The sandpipers nest along the riverbanks and follow the rivers south to winter habitats along the Gulf of Mexico.

Common mergansers follow the streams for food, since they are underwater feeders, traveling south with the approach of winter to find open water. Fish, of course, live in the waters, even the salmon, which is making its way back up the Connecticut River, into which the White flows. Fish attract predators, including both birds and animals, such as the river otter.

Although plants do not migrate annually, they do move in response to changing climate, and riverbanks are among the most common routes.

The Green Mountain National Forest, with the help of the public-spirited Route 100 Lions Clubs, has begun a project to restore and protect the White River Travelway, not only for the unique wildlife living here, but to allow the public access to points along the river to observe the habitat and enjoy its beauty. At various points along the routes of the White River and its tributaries—Hancock Branch, West Branch, and Tweed River— overlooks, boat access, hiking and interpretive trails, and picnic areas have been developed.

Texas Falls is the best known of these, a National Forest site centered around the falls, a short series of cascades and falls rushing, even in summer's low water, through a narrow gorge carved by glacial runoff. A footbridge provides a viewpoint directly over the gorge and a good chance to see the potholes carved by the swirling waters. Unlike many such geological formations, which change direction on their way down, Texas Falls lies in a straight line, so you can see the whole course at once. Like several of the Travelway sites, it has access for wheel-chairs. Above the falls, picnic tables and grills line the river-bank, and upstream is a barrier-free fishing access. Primitive campsites are maintained along the woods road beyond.

Near the junction of State 100 and 125, which is also the junction of the Hancock Branch with the White River, the Hancock Overlook provides a trail through the riverside habitat, as do three other sites along State 100 between Hancock and Rochester. The Riverbend site has a canoe put-in, and the Lions Club Picnic Area in Rochester offers picnic tables. On State 73, a former CCC Camp has a trail, as does the West Branch picnic area. On State 107, farther south, Peavine (the nickname for the winding railroad route that once served the area) picnic area has a trail, canoe put-in, and fishing. Each of the sites mentioned is wheelchair accessible.

For more information on the Travelway or any of the areas that offer access to it, contact the Green Mountain National Forest, Rochester Ranger District, RD 1, Box 108, Rochester, VT 05767; 802-767-4777.

## GIFFORD WOODS

This stand of virgin forest is unusual not only for the size and age of its trees, but also for its location next to a well-traveled road in a major resort area.

*Directions:* On State 100, just north of its intersection with US 4, east of Rutland. The woods are part of Gifford Woods State Park and located across State 100 from its entrance.

Most of the relatively few virgin hardwood forests in the Northeast are hidden far from roads or other human development, little pockets that somehow escaped the ax, the saw, forest fires, and the bulldozer. This one stands beside the road in the highly developed Killington area. Only a few acres between the road and Kent Pond, it is nonetheless almost pristine, with 300-year-old trees over 100 feet tall. Species include ash, beech, birch, maple, bass, and hemlock. It is a climax forest, which means that it has reached the end of the forest progression and that its species can reproduce themselves from seedlings that will thrive in the shade of a deep woods.

It is also outstanding for its variety of ferns—over fifteen species, including maidenhair—and for its spring wildflowers. There are no trails here, and you should avoid walking where others have walked (unlike other places where you should stay on the trail) in order to lessen the risk of compacting the spongy forest floor. The Appalachian Trail goes along the southern border of the woods, and Kent Pond, a popular birding site, lies to the west.

Gifford Woods State Park offers camping in well-spaced, wooded sites. Its stone Administration Building is a fine example of those constructed by the Civilian Conservation Corps (CCC) during the Depression. For more information, contact the park at Killington, VT 05751; 802-775-5354 in the summer or 802-886-2434 in the winter.

# 4

# Western Foothills

**VERMONT WILDFLOWER FARM**

Pioneers in the field (sorry!) of wildflower cultivation for the home gardener, Vermont Wildflower Farm specializes in seeds that can be planted for showy meadow displays.

*Directions:* On US 7, south of Burlington, just south of the turnoff for the ferry to New York.

The flowers you'll see growing in the front along the roadside are not necessarily Vermont wildflowers, but annuals, including bachelor's buttons, anenomes, and other garden favorites. The showy displays are more accurately annual flowers planted in a meadow.

Read the signs to learn—the area needs to be plowed so the seeds can be planted in turned earth. These flowers need to be replanted annually, since only a few plants will reseed and bloom the following year. The results, although not wild, are colorful and certainly brighten the summer roadside.

Native field flowers, with a little help from introduced hardy plants, fill a field behind the building (which includes a theater and a gift shop with a good selection of nature and gardening books). In late summer, the field looks much like the roadsides throughout Vermont, bright with goldenrod and highlights of vetch and jimsonweed, with patches of purple coneflower added for color.

In the woodland behind the meadows, a loop trail passes through a more natural wildflower setting, with signs alongside the trail to identify species and tell a bit about them. Spring is the time to see most of the woodland flowers in bloom, but as late as early September there are some, including the showy red cardinal flower. Be sure to pick up a plant list when you pay your admission fee, so you can match names to the numbered markers.

The farm is open daily, April through late October. Admission is charged. Contact the farm at PO Box 5, Charlotte, VT 05445; 802-425-3931.

## BUTTON BAY

There is enough geological evidence at Button Bay to satisfy the most avid rock hound, but its natural attractions don't end there.

*Directions:* From Vergennes, south of Burlington, follow an unnumbered, but well-labeled route west for six miles, toward Basin Harbor. The entrance to Button Bay State Park is well marked.

Button Bay is best known for the unusual compacted clay buttons that gave it its name. These disks with depressed centers were formed when clay, rich in calcium cement, gathered around the stems of aquatic plants and were buried in later sediment.

---

### HOW CAN SOMETHING SO PRETTY BE A THREAT?

As you stand on the banks of a marshy area, you may notice some beautiful plants with two-foot upright stems covered with pretty magenta purple blossoms that seem to form a carpet. These plants are actually a threat. The plant is *Lythrum salicaria*, purple loosestrife, which likes wet meadows, floodplains, and roadside ditches. It also likes the shorelines of ponds and lakes and any other place that cattails and reeds like to grow. And that is the problem. Although purple loosetrife has been in New England for over a hundred years, its spread has increased over the past several years. It is extremely aggressive and given time it will crowd out native species, killing off cattails, reeds, and other aquatic plants that form an important part of the food chain. Loosestrife itself is not a food or useful resource for wildlife, but its spread weakens the health of wildlife by destroying the sources of their sustenance.

Vermont has begun a program of eradication to slow or halt its spread. If you identify purple loosestrife on your property during June and July, when it first starts to bloom, but before the seeds begin to form, cut the plant off at the base or uproot it. If seeds have begun to form, the plants must be uprooted, placed in plastic bags, and buried in a landfill. Although beautiful in full bloom, this plant is a killer and a great threat to the plant and animal diversity of New England.

---

They continued to be compressed until they formed the buttons found in the bay today.

But the buttons are only a little over 10,000 years old, mere youngsters, geologically speaking. Over 400 million years ago, long before the glaciers deposited the clay from which the buttons formed, the land mass of Vermont was about where Florida is today, and under the waters of a shallow tropical sea. Shells and other sea-floor debris collected and were compressed into limestone when colliding plates of the earth's surface thrust a section of this ocean bottom above sea level. Captured in this were shells of sea creatures caught in the limey mud, forming fossils that can still be found here.

Button Island and Button Point were part of a giant coral reef, containing some of the oldest coral in the world, plus trilobites and sea snails known as gastropods. But while the surface of the earth in other places has been altered by intense heat and other geologic pressures, changing it into igneous rock, the rock of this area has not, so it remained limestone, a sedimentary rock that preserves fossils well.

## Nature Center

In a small cottage at Button Point, the Nature Center is both museum and resource center, with displays of the buttons and fossils, as well as shells, birds, and examples of local natural history. In the Champlain Room, everything relates to the lake, both its history and its nature, including a section on sightings of Champy, the lake's famous sea monster. Books and pamphlets on the lake are available here, too. The center is wheelchair accessible, and handicapped parking and road access are provided. (Other visitors must walk from the campground and picnic area.)

## Naturalist Programs

A park naturalist is stationed at the Nature Center from June through Labor Day, conducting a full schedule of programs for all ages. A typical week's offerings include a discovery walk, a boat trip to the wild gardens of Button Island, a safari in search of nests and other evidence of the creatures that share the park habitat, a walk to find wild foods and medicines, a slide show on endangered plant and animal species in Vermont, a fossil hunt, and a seed hunt. Times for all these programs are posted at the park entrance; all are open to registered campers and day-use visitors alike.

## Walking Trails

The Champlain Trail leaves from the Nature Center and skirts the shoreline of Button Point. Plant life along the trail is different from that found elsewhere, because of the limestone land base. Tiny nodding bluebells bloom in moss-covered rocks, and the rare ram's head lady's slipper blooms in the spring; several species are on the rare and endangered lists. A small mature hardwood forest covers the point.

Watch the flat rocks along the shore side of the trail, after it turns to the right, for excellent examples of gastropod fossils. The best of these are in the rocks forming a long crack, known as a solution cavity, where moving water over a period of time has dissolved the calcium in the rock.

Other trails lead onto the rocky shoreline of the east side of the point, where the shelflike ledges show marks of the glacier moving past, gouging lines into the rock as it moved. Campsites at the park are open and grassy. A boat launch and boat rentals are also available.

The park is open mid-May to Columbus Day, but the nature programs end on Labor Day weekend. For information, contact Button Bay State Park, RD 3, Box 4075, Vergennes, VT 05491; 802-475-2377 in the summer, 802-483-2001 in winter.

## WINONA LAKE

Alive with birds in the morning, the low-lying wetlands around the shallow lake are a fine wildflower habitat as well.

*Directions:* From the divergence of State 17 and 116 at a cross-roads about a mile west of Bristol, go north on the unnumbered road about three miles. On the right you will see a low marshy area and the edge of Winona Lake, and at 3.6 miles a dirt road goes east to a public boat access.

Pond lilies, arrowhead, pickerelweed, and cattails grow out of the water; beside the trail that runs along the northern shore are jewelweed, agrimony, fox grapes, soapwort, chickory, knapweed, Queen Anne's lace, and purple flowering raspberry. The channel through the shrubby marsh to the main lake invites exploration by canoe or kayak. Unfortunately, motorboats are also allowed here, occasionally shattering the peace, drowning out birdsong, and startling waterfowl from their breakfast. But most boaters are quieter, and the noisy ones are at least speedy in their getaway and soon out of earshot.

## ROCKY DALE GARDENS

Take an old hill farm, throw in a couple of California alternative life-stylers, and you end up with one of the finest examples of natural perennial gardening anywhere.

*Directions:* From the center of Bristol take State 116, also called Rocky Dale Road, east about 1.5 miles. The gardens will be on your right.

Bill Pollard and Holly Weir came to Bristol on a visit from California, and in 1980 they settled in, turning a sagging nine-acre farm into an outstanding show garden. Located near Bristol Memorial Forest Park, the farm sits tucked into the steep hill-sides, with a dramatic wall of bedrock rising straight out of the rear gardens.

On arrival, the farm looks somewhat like other nurseries, with rows of potted flowers in the sales area. Go on past these, the house, and the barns, however, and you will encounter the gardens. Molded among lush green pathways, islands and peninsulas of plantings each create a new mood of color, texture, and form.

While each of the outsized islands is different, all exhibit the same thoughtful planning, so that regardless of season there is texture, mass, and color. Anchoring the gardens are the native species such as spruce, hemlock, crabapple, and hazelnut. Another basic element is the massing of different colors of the same species and the grouping of varying species of different heights. The reds, white, and pinks of phlox, soft colors of heathers and astilbes, and bright colors of lilies give life to the spruce, lilac, and dwarf conifers in each setting. Beds of hosta are backed by cliffs and tumbled rock, in whose shelter are shade gardens, fern beds, and dramatically placed sedums and grasses.

Wild is not the right word to describe these gardens, for they are very well tended and carefully planned, but the word does convey the drama of the rocky setting and the scale of the landscape. The gardens are open daily, except Tuesdays, April through October from 9:00 A.M. to 6:00 P.M. Their mailing address is 62 Rocky Dale Road, Bristol, VT 05443, 802-453-2782.

## BRISTOL MEMORIAL FOREST PARK GORGE

Bristol went beyond a stone monument in its memorial to its dead from World War II and Korea. The town built a park that celebrates the beauty of a living stream to symbolize their sacrifice.

*Directions:* From the center of Bristol take the combined State 116/17 east, past Rocky Dale Gardens, to the point where 17 diverges, and follow it about 1.5 additional miles to a parking lot on the south side of the road. The park is well marked.

This waterfall within a park is one of the most attractive and beautiful in the state. While most of Vermont's falls offer, at best, difficult access, this one has well-established and maintained trails, boardwalks, and bridges that provide easy and safe access as well as the best views.

Baldwin Creek makes a series of falls and cascades through a steep and heavily wooded site. From the parking area, the trail leads down past a picnic site along a series of cascades, some of which are as high as six feet. The stream then plummets in a waterfall through a gorge about 15 feet to the base of the falls, where it makes an abrupt right-angle turn to flow through a narrow gorge about 40 feet deep and 120 feet long. At the end of the gorge is a pool, a popular bathing spot, and a number of very large potholes along the lower end of the gorge.

## DEAD CREEK WILDLIFE MANAGEMENT AREA

In the early evening in the summer and during the fall migration seasons, you will meet birders with their glasses, scopes, and cameras, ready to see the wide variety of waterfowl that stops here.

*Directions:* From State 22A, south of Vergennes, go west on State 17 about one mile to the management area headquarters, then two miles to the bridge where 17 crosses Dead Creek. Just past the bridge, turn south onto a dirt road that leads into the area. Park at the boat access at the end of the road.

The slow-moving waters of Dead Creek are slowed even more by the construction of dikes that control water levels to

**In the spirng, wood turtles linger along streams and ponds;
as the weather warms, they search the woods for slugs,
worms, berries, and vegetation**

maintain the most attractive waterfowl habitat possible. Although the best way to see birds is from a canoe, the dike south of State 17 gives an excellent vantage point on dry land.

Over 200 bird varieties have been recorded here, and during the height of the fall migrations there may be as many as 6,000 Canada geese. The access road from State 17 runs along a raised bed that overlooks part of the 100 acres of cattail marsh cut with open channels, where you may see great blue herons or the flash of a red-winged blackbird. Ducks are common, especially mallards and black ducks; overhead you may see marsh hawks or American kestrels. Sandpipers can be seen in the shallow water along the dike at the end of the access road.

Canoe put-ins are plentiful, and from water level you will

see many more birds, as well as the abundant aquatic vegetation. Birds are far less skittish when approached by canoe. Small birds nesting on the edge of the marsh to take advantage of its abundant insect life include several species of swallows, the common yellowthroat, and the swamp sparrow. Canoeists should be very careful from April to June not to approach the shores too closely and disturb nesting birds.

Although the prime attraction of Dead Creek is its bird life, the forests along its edges are notable, too. Look for occasional remnants of the old dry oak-hickory forests that once stood in this part of the valley.

For permission to canoe in the section between State 17 and the dike to its south or to get a map of the entire area, contact the Dead Creek WMA at RFD 1, Box 130, Vergennes, VT 05491; 802-759-2397.

*Of interest nearby:* Two other areas that offer fine canoeing, birding, and views of aquatic plants are located not far south of Dead Creek on marshy streams that enter Lake Champlain. Whitney Creek Marsh and McCuen Slang Waterfowl Area are at the mouth of Whitney Creek, along State 125, about two miles south of Chimney Point. A public boat-launch area offers access to the lake and to the creek and marshlands.

## SILVER LAKE

Once the site of religious camp meetings and a large hotel, the land around the lake has returned to nature, a place of intimate contact with the land.

*Directions:* Follow State 53 to Branbury State Park. Although there is parking on the east side of 53 about .2 mile south of the park, vandalism to parked cars there makes parking at the state park preferable, even if the walk is longer. The trail is on the east side of the road and is marked for the Falls of Lana. It is also reached by taking the Goshen-Ripton Road, then Silver Lake Road to a parking area at the end of Forest Road 27. Leiscester Hollow Trail goes there from State 73, a 3.5-mile hike.

Although the area was originally settled in the first quarter of the nineteenth century, it wasn't until 1879 that a missionary from Montreal named Frank Chandler decided that this was an ideal place for a religious camp meeting. In that age of religious fervor, the camp became popular, and roads and a sixty-room hotel were built to accommodate the customers. In 1911, water rights were sold to a power-generating company that still uses the waters of the lake today for the generation of electricity. By 1919, the hotel was closed, and over the years it, and all of the buildings, have disappeared except for their foundations.

Sitting 670 feet above Lake Dunmore, Silver Lake is a 1.6-mile hike from State 53, much of the initial part following a steep gravel mountain road. The lake is a mile long, but never more than a quarter mile wide. It sits at a juncture of hard Cheshire quartzite on the west side and Forest Dale dolomite to the east, the quartzite forming Chandler Ridge along the western shore. On the north end of the lake there is a sandy shore with a picnic area.

The highlight of the lake, however, is its interpretive trail, a 2.5-mile hike around the lake with thirty-two explanatory stations that detail its natural and man-made features. An out-standing trail map and pamphlet are available from the Forest Service visitors center at the campground on Lake Dunmore. Trees found along the trail include white pine, paper (white) birch, yellow birch, ash, and beech. Other varieties of plants highlighted are American hornbeam, serviceberry (shadberry or smooth Juneberry), wintergreen, tree club mosses, lily of the valley, partridgeberry, common polypody fern, and field horsetail, a primitive plant that has hard mineral silica within its cell walls. Foundations of the hotel and its barn can also be seen.

There are seventeen free primitive campsites at the lake, but remember that these are carry-in and carry-out. Fire rings and hand water pumps are available. No motor vehicles, or motors of any sort, are allowed in the area, except for snowmobiles, which are permitted during the winter months. Contact the park at RFD 2, Box 2421, Brandon, VT 05733; 802-247-5925.

## EDGEVIEW ANTIQUE ROSE GARDEN

A labor of love, this unique garden brings back the roses of the past. Wander its grassy paths and enjoy the sight and fragrance of roses that charmed the Belle Epoque.

*Directions:* Take US 7 or State 73 to Brandon, which is about halfway between Rutland and Middlebury. From the center of town, follow Park Street easterly and take a left onto Marble Street, which is also State 73 east, just on the outskirts of town. The garden is at number 27.

When Albert and Shirley Hill retired from teaching, they were finally able to satisfy a lifelong ambition. Selecting only roses that were cultivated prior to 1920, they have created a garden that is unique in Vermont and one of only a handful in the United States.

Opening in 1990 with fifty antique varieties, the garden now grows more than 200 roses. The collection includes the common old "pasture" rose, rosa Virginiana, the very fragrant Blanc double de Coubert (a white midseason bloomer), and Rambling Rector, which can grow into a jungle over twenty feet high.

Varieties on display include single (five-petaled) blossoms as well as double (with upward of twenty or more petals) blossoms with foliage as varied as the deep green of the rugosa varieties to the mossy green and hairy surfaces of the Henri Martin, which produces a dark red bloom of over sixty-five petals. The roses bloom from early June to the end of the summer, and some may be blooming as early as the last week of May. This is a show garden for viewing, a place to see plants that you are considering adding to your own garden. They don't sell plants here, but if you ask, Albert will help you find a source for the rose that you fall in love with. There is an admission charge of $3.50.

## BRANDON GAP

One wall is Mount Horrid, whose Great Cliff hangs 700 feet above the road. A short (.6-mile) but very steep hike on the Long Trail leads to the lookout on the cliff.

*Directions:* Follow State 73 northeast from Brandon. The road runs uphill for several miles, more steeply as it nears the gap. At the top, there are parking areas for climbers or continue downhill a short distance to a pullout looking out onto the mountain.

Even if you want to climb the Mount Horrid trail, go on down to the parking area on the northern side of the road so that you can fully appreciate the cliffs. Across the valley the mountain rises to sheer rock cliffs while below beavers create a pond and marshy environment that contrasts with the starkness of the cliffs.

The trail leaves from the top of the gap, traveling north along Mount Horrid to the Great Cliff, 700 feet above the roadway. This is about a .6-mile hike along a part of the Long Trail. The climb is moderate to difficult, but the reward is magnificent views over the valley and beyond. Hiking to the cliffs will

probably be limited during the spring to protect pairs of peregrine falcons that nest on the sheer face of the cliff. The cliffs are
precipitous and can be dangerous, so you should wear appro-
priate footgear.

## BRANDON HOGBACKS

If you were traveling along a road in Europe, you'd recognize
these pillars as the ruins of an old castle, but in Vermont, and
overhanging the rim of a gravel pit, you aren't quite sure what
they are.

***Directions:***  From Brandon take Pearl Street west over the bridge
and look for the gravel pit below the road on the left, about half
a mile from the bridge. To see the pillars, you will have to look
back, since they are hidden from view by a stand of sumac
foliage as you approach from the east. Approaching town from
the west, they are visible from the road on the right just over half
a mile from the covered bridge.

Brandon is located on what geologists call a terrace, an
ancient beach on the shores of Lake Vermont (see the box "Lake
Vermont" in section 1), where gravel, sand, clay, small stones,
and pebbles accumulated. In places, gravel was compressed into
a cementlike substance, with larger rocks embedded throughout
it. The deep terrace accumulation makes a fine source of sand
and gravel, a great amount of which has been removed, creating the pit you see. As the pit was excavated, these areas of
cemented gravel were left behind by the workers, who simply
moved on, removing softer material. The pillars are the exposed
and weathered remains of these cemented gravel columns.

## SHAW MOUNTAIN NATURAL AREA

One of the most diverse natural habitats in the state, Shaw Mountain has a shrub swamp, an oak-hickory forest, a vernal pool, and a community of calcium-loving plants, as well as twenty-five rare plant species.

*Directions:* From State 22A, north of Fair Haven, go west on an unnumbered road to Benson. Continue through the crossroads in the center of town and in one mile go left on a gravel road. Take the first right at .6 mile and continue 1.5 miles to a small pullout on the left and a small sign identifying the trailhead.

An uplift of limestone rising about 500 feet above the surrounding landscape, Shaw Mountain's rocky slopes and ledges ascend steeply from a marsh community that extends from the northern shore of Root Pond. The trail (which is not a labeled interpretive trail, but one where you must spot the natural attractions yourself) begins along the edge of this marsh and quickly rises along the hillside facing Shaw Mountain. It passes through a particularly attractive woodland where the forest floor is completely grass covered and vines form interesting patterns against the green. Near the sign-in box is a black birch tree, rare here (and labeled). The trail descends to the marsh, which it crosses on a boardwalk. Birders may decide to stay right here.

A short distance up the slope of Shaw Mountain, the trail bears right to make a loop around the summit. At this point, you should check the back side of the tree where the sign is. When we were there, there was no sign on the back of the tree to alert returning hikers of the intersection, and the open forest floor does not make trails easy to spot, especially if you are busy looking at the flora and fauna. We suggest that you mark this point in some way to avoid beginning the circular route a second time;

we have begun carrying a long piece of colored ribbon for situations like this.

As you climb to the western rim of the mountain, you will pass through fine stands of fern and along the edge of a steep ravine. The trail blazes are difficult to spot in the light-dappled woods, but the trail circles the rounded top of the mountain and returns to the marshy lowland.

In the spring you may see three trillium species – unusual to find in close proximity because of their differing soil preferences – as well as bloodroot, round-lobed hepatica, and wild ginger. The flowers of the latter are hard to spot, but its deeply cut, almost heart-shaped leaves grow in colonies in moist, shaded spots. Walking ferns can be found on the calcium-rich limestone outcrops. Among the rare and endangered plants are snowy aster, lanceolate cress, squawroot, three-lobed violet, and four-leaved milkweed.

Allow two hours for a leisurely look at the various environments. The area is protected by the Nature Conservancy, 27 State Street, Montpelier, VT 05602; 802-229-4425.

## MEADOWSWEET HERBS

A terrace of herbs labeled for identification and an ambitious Renaissance garden are only two of the many reasons herb lovers gather here.

*Directions:* From State 103 in Cuttingsville, take an unnumbered road north to Shrewsbury; turn right to North Shrewsbury, about six miles from State 103. Make another sharp right before the village store (a sign points the way) and continue one mile to the farm, which sits astride the road.

As Polly Haynes builds the Renaissance garden and the multitude of historical gardens she has planned for Meadowsweet, those interested in authentic historic restoration gardens watch carefully. The plan is historically accurate, based on a design of William Lawson's 1688 garden. Within the large grid plan, each separate garden – Bible, tea, Colonial kitchen, posy, Victorian, and even a garden for sagging spirits – will be enclosed as though it were a room, separated by gates. The process of watching these gardens grow is as interesting as the gardens themselves, and very instructive to those who hope to create their own garden to match a period home.

Greenhouses are filled with plants for sale, and a bright, well-stocked shop has a good assortment of herbs, food items, and books, as well as potpourri and herbal gifts.

Open 10:00 A.M. to 5:00 P.M., daily from May through October, and on weekends only from November through April. Ask to be put on the mailing list for programs, shows, and classes. 729 Mount Holly Road, North Shrewsbury, VT 05738; 802-492-3565.

## CLARENDON GORGE AND FALLS

This deep ravine with fast-flowing water, cascades, and pools has an entrance that would thrill Indiana Jones.

*Directions:* Traveling northward on State 103 from Shrewsbury to Clarendon, the road makes a broad turn at a railroad crossing. The parking lot, for the Appalachian and Long Trails, is on the left, or southwestern, side of the road. Watch traffic carefully entering or leaving this lot.

The path to the gorge is at the southern end of the parking area, close to the railroad crossing sign. (The major path in the center of the lot leads to the Long Trail, bypassing the gorge.) Follow the path downhill to the suspension bridge. If you are uneasy with heights, we suggest only one person cross the bridge at a time to cut the swing sensation. Straight ahead is a path down to the riverbanks through a little clearing. Follow the river back under the bridge to see potholes and water-worn rock formations.

The best view of the gorge is from the swaying footbridge. While the rock walls of the ravine are only about thirty feet deep, very steep wooded hillsides continue their precipitous climb, creating walls from 100 to 150 feet tall. The river itself flows from an alluvial valley, making a sudden right-angle turn. At this spot there was a segment of softer rock between the hard rock of two neighboring hills. The river took advantage of this and wore its way through, creating the gorge and making this one of the best examples of differential erosion of hard and soft rocks in the state.

Soft rock and swift water also encourage the formation of potholes, and they are found here in abundance. The stream, which is filled with huge boulders, passes through the gorge forming pools and small falls or cascades of under three feet in height. Botanists will like this site for its great variety of mosses, liverworts, and vascular plants.


## CLARENDON SPRINGS

As you drive into the village, the well-kept facade of the hotel looms grandly ahead, and you want to rush in and reserve a room.

*Directions:* From State 133, southwest of Rutland, about two miles south of its interchange with US 4, take an unnumbered road to the east. Clarendon Springs is about a mile from the intersection.

Except for the absence of people, the village of Clarendon Springs looks much as it did in its heyday before the Civil War—

**Saw-whet owls live among evergreen forests, hunting for rodents at night**

manicured lawns, tall shade trees, tidy homes, and a general store. Overlooking it all is the brick Clarendon House, its triple pillared porches wrapping around three sides. It is imposing and inviting now, just as it was when the gentry traveled from Southern cities to summer here and take the waters. The hotel building is real, but the rest is an illusion, for the interior of the building has not been repaired since it closed as a hotel early in the century. It is now used to store antiques, but its owner has taken meticulous care of the exterior and preserved the structure and its fine old white wooden porches.

The lure of Clarendon Springs was, apart from its lovely setting and the supposed health-restoring properties of its water, the belief that its waters increased fertility. The spa was the darling of the Southern aristocracy. All this ended abruptly with the Civil War, and the hotel continued to decline along with the popularity of mineral springs.

What of the springs? You can visit them and try the water, which tastes slightly of minerals. Park in the center of the village and walk downhill behind the hotel to the river. Cross on the footbridge and follow the trail to the left, through a small marshy meadow to the shingled Victorian springhouse. The spring has been piped underneath the building, and you will see it easily. We are not responsible for the results of your testing the water.

*Of interest nearby:* Another spa resort popular later in the nineteenth century was at Middletown Springs, a few miles farther south on State 133. Unlike Clarendon, its springs were not mineral, just pure fresh springwater. Its Montvert Hotel is long gone, but the springhouse has been restored along with a small park on the hotel grounds beside the river. Interpretive signs describe the springs and the hotel. Visitors can sample some of the grandeur of that era by staying at Middletown Springs Inn, on the green a short walk away. Originally the home of a prosperous local family, the Italianate building has retained the historical features that made it the showplace of this fashionable spa town. Contact the Inn at Box 1068, Middletown Springs, VT 05757; 802-235-2198.


## BIG TREES NATURE TRAIL AT LAKE ST. CATHERINE

The area around the lake was farmland well into this century, but the trees along the ridge were not cut to clear fields. Today they provide mature examples of a number of native tree species.

*Directions:* The park entrance is on the west side of State 30, south of Poultney.

Lake St. Catherine State Park is best known for its clear lake and fine beach, as well as its campground, popular with those who enjoy boating on the lake. But close to the entrance, a trail of about a third of a mile leads past a fine collection of hardwoods. An interpretive brochure, available at the park entrance, identifies the major species and gives a fascinating view of how the wood, bark, nuts, and even the sap have been used by humans.

While most visitors are aware of the syrup made from the maple, many are surprised to learn that a tree can be almost entirely hollow, but still be healthy and produce its full complement of syrup each year. The sap of other trees was used by the Indians for sweeteners: white ash makes a bitter dark sugar, and the butternut produces both syrup and nuts. Hickory wood was used to make baskets, chairbacks, and gunstocks, while the black cherry produced a cordial and basswood a fine, strong fiber.

These old-growth trees may have survived clearing and timbering because of their location along the ridge or because they had other value, such as the nuts from the butternut and hickory trees.

The sixty-one tent and lean-to sites in the campground are wooded and well spaced, some overlooking the shore. Administration at this park is enhanced by the enthusiastic Youth Conservation Corps members who often staff the contact station and work on other projects at the park. Unlike several other parks, this one is open until Columbus Day weekend. For information, contact the park at RD 2, Poultney, VT 05764; 802-287-9158 in the summer or 802-483-2001 in the winter.

## SLATE MINES

Startling as they appear suddenly beside the road, the abandoned quarry hole is now filled with water, and the giant hills of mine tailing have trees growing from their summits.

*Directions:* On the north side of the village of West Pawlet, on both sides of State 153, northwest of Manchester.

Although a more "formal" historic trail of a slate town exists in Bomoseen State Park, farther north, the slate quarry at West Pawlet lies adjacent to the road, visible to everyone who enters or leaves the town from the north side. The steep walls of the quarry, filled now with water, appear suddenly as a bridge crosses the water that now fills it. Hills of slate scrap—like granite, as much as 85 percent of slate mined might be dumped as waste—line the roadside, held in place by walls of stacked slate.

Slate was formed here over 500 million years ago when a sea covered the area. The sediment of soft clays and organic debris settled in layers, packed hard, and then were thrust upward. The pressure and heat that were exerted on the hardened sediments further compacted them, creating a metamorphic stone with a distinct layering. The thin sheets into which slate could be split made it useful for roofing, floors, and a facing for mantels and fireplaces that could substitute for more costly marble.

The first clue you have to the origins of West Pawlet's compact nineteenth-century downtown are the roofs—nearly every one of them made of slate. As you travel through the surrounding area, look for patterned slate roofs on homes and farms. Slate has several different colors, and these were often arranged in elegant designs on the roofs of Victorian homes.

*Of interest nearby:* In November, the Pawlet Fire Department holds one of the most popular of Vermont's famous game suppers.

The food is prepared by local residents, most of it from wild game that has been hit by automobiles and held in freezers by the state. A typical meal might include roasted venison, braised squirrel or rabbit, moose meatballs, pot roast of bear, a game stew, and accompanying vegetables and salad. If you can still think of food after this, there's a wide variety of homemade pies. You sit at long tables in the firehouse, where you meet both local residents and people who travel some distance for this annual feast. The price is under $10. The supper begins at 5:30, but the line forms much earlier. The supper is always on the first day of deer hunting season, normally the second Saturday in November; watch local newspapers or bulletin boards.

## MERCK FOREST AND FARMLAND CENTER

On 2,800 acres of land set aside in 1950 by George Merck, the center includes 28 miles of hiking trails, hike-in camping, and cabins, as well as a full schedule of nature programs.

*Directions:* From State 30, north of Dorset, take State 315 west at the village of East Rupert. The entrance to the Merck Center is on the south side of State 315, between East Rupert and Rupert. In the winter, when the access road is not plowed, you can park on 315 and ski to the center.

Environmental and educational workshops, a small demonstration farm, a manageable mountain to climb, a working sugar bush, year-round camping, and primitive log cabins deep in the woods are only a few of the activities and facilities at the Merck. The visitors center is photovoltaic, not even hooked into a back-up power source.

Education for young people is a primary goal here. "Kids are more connected today with a rain forest 3,000 miles away than to their own backyards; we're trying to get them reconnected here," says the center's director, Richard Thompson-Tucker. They do this through school visits and a year-round series of unique programs that attract families. At the end of November, for example, is the Full Moon Hike, and in December visitors can harvest their own Christmas trees and ride back behind the center's team of Belgian horses. Year-round "camping" is offered in cabins, where wood is supplied for the woodstoves, but bedding and other gear must be carried in on foot (or on skis). Primitive shelters and tent sites are sprinkled throughout the property.

### The Discovery Trail

A self-guided nature trail leads from the barns to the visitors center, a distance of about three quarters of a mile. At ten stops along the way, the features of the land and the plants and animals it supports are highlighted, with emphasis on how all living things connect to each other. Subjects discussed in the interpretive brochure range from the decomposition of a log and the make-up of forest soil to the role birds play in the life of a forest.

For reservations to camp there or information on upcoming programs, contact the Merck Center, Route 315, Rupert Mountain Road, P.O. Box 86, Rupert, VT 05768; 802-394-7836.

# 5

# Upper Valley

## GLEN FALLS

Two falls are the reward for this short hike in a pine-and-hemlock forest.

*Directions:* From I-91, take exit 15 to US 5, south of Fairlee. Take Lake Morey Road, passing between the Lake Morey Inn and the golf course, and follow the road along the west shore of the lake. About a third of the way up the lake is a public boat access on the right. Park here and walk back along the road to the trail, which leaves the road opposite a tennis court.

Most of the trail is easy walking on a moderate grade. The last part is along a steep and slippery bank, but you can see the falls from the stream without climbing. It is about a ten-minute walk to the upper falls. The trail follows a narrow stream through a hemlock forest, and the valley quickly narrows to a gorge as the first falls appears. Continue along the trail, which will suddenly begin to climb steeply along the edge of the upper gorge.

The upper part is about 100 feet long and 40 feet wide. Through this dramatic break in the rock the stream falls 25 feet into a pool at the base of the fissure. The jagged sides of the gorge are nearly vertical and the path passes quite close to the edge in places. The soil is covered with pine and hemlock needles and can be extremely slippery. Avoid the edges of the gorge as there is nothing to prevent a fall.

## ELY COPPER SMELTER AND MINE

A long-abandoned copper-smelting operation provides interesting historical perspectives and a look at some fascinating minerals as well.

*Directions:* From State 113 in West Fairlee (not to be confused with West Fairlee Center) take the road toward South Vershire for slightly over one mile. The road is called Copperfield Road because it once led to the settlement of Copperfield, now abandoned and mostly disappeared. At the top of a rise the road comes out onto a rather desolate level area. You can park at the pullout on the south side of the road.

One look at the deep-rust-colored river shows that iron, as well as copper, is present in the rocks of this area. All around the site are huge piles of the slag generated by the copper-smelting operation that operated here. Some of the slag pieces have interesting and tortured shapes and colors.

Take some time to wander around and see the remains of the furnace and other buildings that supported the plant. For an interesting view, and a look at how they brought the ore from the mine to the smelter along an earth ramp, climb up to the top of the huge granite retaining wall along the rear of the site. At the west end, a dirt road, overgrown to a path, leads through the woods for about three-quarters of a mile to the mine dumps.

Garnet, hornblende, calcite, and malachite can be found among the rocks.


## OLD CITY FALLS RAVINE

Upper and lower waterfalls are set in the midst of a wild ravine, whose rocky walls narrow to form a gorge at the falls.

*Directions:* From State 132 at South Strafford, take the unnumbered road north to Strafford. In the village, take Harris Road, which forks to the right beside the large church on the hill. In about half a mile bear left. In a short distance the road descends a short hill, crosses a bridge, and starts to rise. Partway up the hill, on the left, is the entrance to Old City Park, marked with a sign. The trail begins at the far end of the parking area.

After a gentle course through a hemlock forest, the trail makes a switchback turn and descends steeply down the embankment of a ravine to the wide boulder-strewn stream bed below. Following Old City Brook upstream, as it cascades over the granite boulders, you catch a glimpse of the upper falls, through the trees, high above. The stream races through a steep narrow rock chute about thirty feet long and drops into an inviting pool at its foot. Above, it falls in a steep cascade, almost a falls. A narrow path along the cliff leads to a view of the upper falls, where the stream plummets about twenty feet through a fissure in the rock into a pool. Parts of the trail, particularly that section that leads to the base of the upper falls, are slippery and dangerous and proper footwear should be worn. The hike takes about forty-five minutes round-trip.

The well-kept park at the trailhead has picnic facilities, grills, and a rain shelter. No admission is charged.

## QUECHEE GORGE

The small Ottauquechee River has cut a deep cleft in the rock, creating one of Vermont's best-known attractions.

*Directions:* US 4, west of White River Junction and east of Woodstock, passes over the gorge. Parking is plentiful at either side of the bridge. On the east side, a trail leads northward along the gorge to an overlook; another trail goes south, under the bridge and along the gorge.

Over 13,000 years ago, the Wisconsin glacier began to melt and its sand-, gravel-, and boulder-laden waters started to course over the underlying metamorphic rock, acting as a slurry and grinding a wide channel through it, eventually creating a gorge 160 feet deep and nearly three-quarters of a mile long.

The best overall view is from the highway bridge. Sidewalks on either side give a safe viewpoint from which you can see the length and depth of the watercourse. Beware of the traffic on this busy highway and use the underpass on the east end of the bridge to cross the road. The sides of the gorge are vertical. Climbing the walls of the gorge is not only forbidden, it is suicidal.

The trail to the bottom of the gorge runs south along the edge of the cliff and then continues on a moderate grade to the bottom of the cut. Part of the way the path is paved with crushed stone. Just beyond the point where the crushed stone ends, an overlook gives a good view over the river. This point has the best views of a series of small falls and potholes in the river below. The trail is fenced along its entire length, and it is unwise to even think about climbing it to get to the edge. At the bottom the trail ends at the river as it reaches a broad turn and slows, passing over a rocky bottom. It is possible to go upstream a few feet, but the sheer sides of the chasm bar access into the deeper parts of the gorge. Two trails lead from the gorge trail to the Quechee State Park camping area (thirty campsites). These trails are restricted to use by campers only.

The trail from a small park north of the highway also passes directly along the side of the gorge. While there is some opportunity to look over the edge, the real purpose of this short hike is to see the dam and falls upstream, where the river drops out of Dewey Mill Pond. You get an oblique view of the falls from a small hyrdroelectric plant at the end of the trail. A dam has partially obstructed the falls, which must have been a splendid sight. Unfortunately, sumac has been allowed to cover the best views of the remaining falls.

An explanatory pamphlet, with a trail map and diagrams illustrating the creation of the gorge, is available at the information station on the east side of the gorge. There is no admission fee. For more information, contact Quechee State Park, Quechee, VT 05059; 802-295-2990.

## VERMONT INSTITUTE OF NATURAL SCIENCES

A living museum of birds of prey and two well-documented nature trails are featured at the headquarters of this nonprofit organization dedicated to environmental education and natural-history research.

*Directions:* From US 4 in Woodstock, take Church Hill Road, which leaves the west end of the green opposite the Town Hall. The institute headquarters is about 1.5 miles on the right. You can also reach it from State 106 by taking Church Hill Road beside Kendron Valley Inn in South Woodstock.

Founded in 1972, the institute provides support for education in schools and to the public at large, as well as programs for the protection and preservation of endangered species. The center has a library of over 5,000 volumes on natural history and the environment and the Hoisington Natural History Slide Collection of over 35,000 slides. The visitors center has a gift shop with a very good selection of natural history and environmental books and materials for both adults and children.

A wide variety of special programs is available throughout the year. Some are held at the institute's center, and others are field trips held throughout the state. They have included birding trips in the Champlain Valley, mushroom foraging in the Connecticut Valley, a snowshoe-building course, and Day Camp field trips for children in the autumn. Brochures of the programs are available.

### Vermont Raptor Center

Originally conceived as a veterinary facility for injured birds of prey, the Raptor Center has grown into a living museum of flight habitats for twenty-six varieties of raptors. Injured birds are brought here from all over the state for treatment, rehabilitation, and release to the wild. Raptors whose injuries are too serious for later release are treated and housed at the center for research and public education. Over forty live birds are usually on view, from the tiny saw-whet owl to bald eagles and peregrine falcons, creating a unique opportunity for up-close observation of birds usually seen only from a distance.

The center is open from 10:00 A.M. to 4:00 P.M. daily year-round, but is closed on Sundays from November through April. Admission is charged. Information on the Raptor Center, the Bragdon Nature Preserve, and all programs is available from the Vermont Institute of Natural Sciences, Woodstock, VT 05091; 802-457-2779.

## Bragdon Nature Preserve

Fanning out below the Raptor Center, the two trails of the preserve offer a variety of environments. Except for a few brief steep or slippery sections, the trails are easy to walk. Booklets describing each trail can be borrowed at the trailhead box or purchased at the reception desk. These are well written and illustrated with line drawings of the birds, insects, trees, and plants found along the way; each has a detailed trail map. There is no admission fee for the nature trails.

The Interrelationships Trail shows how plants, animals, and the earth depend on one another. Each of the twenty-five stations has been selected to illustrate some facet of these relationships, from the study of insect galls, plant succession, and the biological uses of dead trees to the complex balance of wetlands.

The Communities Trail explores how an interdependent group of plants and animals use available resources and how they have been affected by people. Twenty-four stations illustrate the communities present in a pond, on and along a stone wall, in dead trees, and in a pine forest. Views of Shrewsbury and Killington Peaks appear from station 9 while Mount Ascutney, to the south, is seen from station 16. Station 12 has a rope "rail" where visitors are invited to close their eyes and depend on the senses of smell, hearing, and touch while using the rope as a guide through the woods.

## ESHQUA BOG NATURAL AREA

A wetland fen, the bog contains holdover species from the period of the last glaciers.

*Directions:* On the east side of Woodstock, US 4 makes a sharp 90-degree curve. Traveling from the west (from the village),

Hartland Hill Road leaves US 4 directly ahead; from the east, the road will be a sharp turn to the left. Follow Hartland Hill Road for about 1.2 miles. Bear right at the fork, on Garvin Hill Road, for an additional 1.2 miles until you find a narrow pullout on the right. Be careful of mud here in wet weather. The trail starts at a sign a few feet farther along the road.

The trail along the periphery of the fen is short and easy, and a boardwalk allows access across the center of the bog. Ten marked stations indicate the highlights of the bog, which is particularly well known for its orchids, including white and green bog orchids and pink, yellow, and showy lady's slippers, which bloom in profusion in the spring. Ground plants include partridgeberry, goldthread, and bunchberry. Tree varieties represented are larch (tamarack), maple, and pine. Labrador tea, pitcher plant, marsh marigold, sensitive fern, sundew, beaked hazelnut, witch hazel, bloodroot, painted trillium, cattail, and other wetland species can be easily identified. Access to this bog is particularly easy because of the boardwalk. A trail map is available at the trailhead box, but a wild-plant guide would be useful since not all species are clearly identified.

Information on the bog is available from The Nature Conservancy, 27 State Street, Montpelier, VT 05602; 802-229-4425, or from the New England Wild Flower Society, 180 Hemenway Road, Framingham, MA 01701; 508-877-7630.

## SPRINGWEATHER NATURE AREA

The waters of an impounded lake are fairly shallow and are a fine habitat for fish and waterfowl, including blue herons.

*Directions:* North of Springfield on State 106 about 2.5 miles, take Reservoir Road north a little over 1.5 miles. The nature area will be on the left, or west, side of the road. Visibility as you turn in and out is limited, so be careful.

This natural area, a joint project of the Ascutney Mountain Audubon Society and the U.S. Army Corps of Engineers, has been designed for environmental studies by school groups and individuals. The lake below was formed by the damming of the North Branch of the Black River and smaller streams in a flood-control project.

The three series of trails, blazed in red, blue, and green, each of which has several side-trip alternatives, are well marked and are not difficult, taking a half hour to an hour depending on how many stops and alternative trails one takes along the way. Once a farm, the property has good examples of succession growth and a number of different habitats: forest, field, small pond, flood plain, and shallow lake. This diversity supports a wide variety of plant and animal life. Nodding, white, and painted trillium, jack-in-the-pulpit, wild ginger, and several varieties of fern, including fiddleheads, can be found here. Herons, common mergansers, black and mallard ducks, and Canada geese use the property at various times of the year, as do chickadees, downy and hairy woodpeckers, nuthatches, and jays.

Mount Ascutney, of which there are several good views from the trails, is a good example of a monadnock, a mountain formed when molten rock was forced up into a large cavity in the earth's crust. When the softer and older rock above the mound was worn away by glacial action, the remaining harder new rock was left, forming the mountain standing alone in the scraped-down landscape.

A trail guide with a map is available at the information board at the trailhead across from the main parking lot. There is no admission fee.

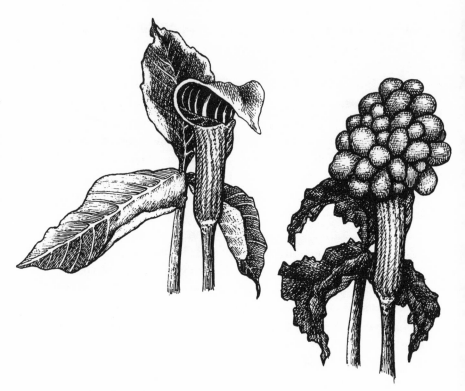

**Jack-in-the-pulpits bloom in the early spring and follow in the fall with clusters of bright red berries**

### SPHAGNUM MOSS

Sphagnum moss grows in wet places with little or no water flowage. It likes acidic waters, and for that reason it is most often found in bogs. This plant is unusual in that it has many nonliving cells spaced among its living chlorophyll-bearing cells, which allow it to store up to twenty times its own weight in water. Used by man for things as diverse as packing live plants for shipping and as a soil conditioner in gardens, sphagnum also forms peat. Ancient compressed peat was (and still is in many places) mined from bogs and used as a fuel for cooking and for heat.

## NORTH SPRINGFIELD BOG

This small bog, hidden away in the most unlikely place at the top of a gravel pit, is readily accessible and has some of the finest examples of bog flora in the state.

*Directions:* Follow State 106 north from the center of Springfield until State 10 joins on the left. Take 10 for a few hundred feet to Fairgrounds Road, which heads north just beyond the Riverside Junior High School. In about 2.5 miles you will come to a gravel pit on the west side of the road. Park in the first area, by the sign designating free sand for residents to use in winter. The trail is an old road to the left partially blocked by a barricade of dirt and rubble.

An open gravel road growing with weeds leads gently uphill on an easy curve to a junction with another abandoned road, a distance of only a few hundred feet. Look to the right for a sign at the entrance to the bog. A floating boardwalk leads into its center.

Barely more than an acre, the little bowl is surrounded by a forest of white pine, maple, birch, and hemlock. Scientists have estimated that the bog is over 10,000 years old. In mid-September the pitcher plants turn bright red, and you can sit down and peer under the branches of the shrubs into the beautifully veined leaves and watch them have lunch. Here we found one of the most enormous pitcher plants we have ever seen, with pitchers over a foot tall. The bog also has a wide variety of wetlands plants, including several varieties of ferns.

The "Superdeck" boardwalk floats on the top of the bog mat, so you have the sensation of walking on the bog even while your feet stay dry. As you walk along, watch the plants and hummocks around you jiggle as you set the mat to quaking. Stay on the walk because not only is the mat dangerous, but it is fragile as well.

The protected area is a joint project of the town of Springfield, the Ascutney Mountain Audubon Society, the Vermont Agency of Natural Resources, and the National Park Service. For further information, call 802-885-2779.

## THE HARTNESS TELESCOPE

The Hartness telescope became the focal point for a group of amateur astronomers, whose interest was so strong that it eventually led to the building of the Mount Palomar Observatory.

*Directions:* From I-91, take exit 7 and follow State 11 to Springfield, where it becomes Main Street. At a traffic light in the center of town, turn right onto State 143 and make an almost immediate left turn up the hill, following the signs for Hartness House.

James Hartness was one of those prodigiously productive men of the turn-of-the-century whose inquisitiveness and drive helped to change the world. Inventor of the turret lathe, between 1886 and 1933 he claimed over 120 machine patents. Among his avocations, astronomy was perhaps the strongest. Dissatisfied with telescopes available at the time, in 1910 he designed and built one of his own. This 600-power turret equatorial telescope was an outstanding accomplishment for its time. It has a 10-inch lens with a focal length of 150 inches and is of the descoude, or elbow, type in which a prism at the base of the tube bends the light 90 degrees to the eyepiece. He gathered around himself a number of similarly interested amateurs whose number included Russell Porter. Porter later became a chief founder of the Stellafane Society of amateur astronomers and, with the society, was instrumental in the construction of the Mount Palomar telescope in the 1930s.

The Hartness telescope is capable of focusing on any object visible in the sky and due to its unique design can be used at all times of year, including the dead of winter. In order to ensure comfort, Hartness had a 250-foot tunnel built to the telescope, as well as a suite of underground studies and workrooms, all heated and well ventilated. These rooms are now the museum of the Stellafane Society. The museum and telescope are not open at all times, but tours are conducted at 6:00 P.M. daily, except Sundays, year-round. Although the telescope and museum are situated in the Hartness House Inn, nonguests are welcome to join in the tours. Contact the Hartness House Inn, 30 Orchard Street, Springfield, VT 05156, 802-885-2115, for tour reservations.

## BLACK RIVER FALLS

Although the falls are somewhat marred by the presence of a dam, their height and presence in the middle of the town of Springfield make them worth a stop.

*Directions:* Park near the junction of State 11 and 106 with 143 and walk to the Park Street bridge, nearby. The bridge offers the best view of the falls.

The old Indian name for the falls was "Comtu," which refers to the great noise generated as the river drops over its rocky chasm 110 feet in the course of only an eighth of a mile. The power of this river was the energy source that supported the early industries of Springfield, generating as well the wealth that led to the town's growth and prosperity. Today the river still provides power, but now it's in the form of the electricity produced by the hydroelectric plant visible along the shore.

# 6

# Southern Vermont

## EQUINOX PRESERVATION TRUST

The Equinox, whose imposing white facade and columns have been the centerpiece of Manchester Village since the early 1800s, has recently placed 900 acres of land on the eastern slopes of Mount Equinox in a conservation trust status. About 25 acres of the woodland is home to endangered plant species rare in Vermont; ski and hiking trails give public access to the area.

In cooperation with the Vermont Institute of Natural Sciences in Woodstock (see Section 5), the land will be used as a teaching area for outings, nature walks, and other programs guided by VINS naturalists. These programs continue on a regular schedule throughout the year and cover native wildflowers, edible plants, birds, insects, fish, weather, minerals, mushrooms, beavers, wolves, astronomy, and other topics. Many programs combine other interests, such as the aerobic walks led by the Equinox Spa director and a naturalist, and art classes where children learn the features of animals by making masks of their faces or where adults learn how to draw birds from nature or photograph flowers in the Hildene gardens.

Geared to the seasons and to hands-on experiences, typical programs have included identifying hawks during migration, gathering wild greens and tubers, listening to bats with a "bat detector," preparing wild vegetables with the Equinox chef, and climbing to the rare plant colonies on the lime-rich mountain outcrops.

All programs are open to the public; a small fee is charged. For schedules and reservations, contact the Vermont Institute of Natural Sciences office at The Equinox, Manchester Village, VT 05254; 802-362-4374.

## BOSWELL BOTANY TRAIL

In addition to the native wildflowers already growing in the woods adjoining the Southern Vermont Art Center, specimens of other native plants have been introduced to create an unusual collection.

*Directions:* From US 7A in Manchester Village, take West Road, which leaves from the west side of 7A just north of The Equinox. The entrance road to the center is signposted on the left. West Road, incidentally, is a good way to avoid the massive congestion of Manchester Center if you want to go west on State 30.

Over thirty varieties of native ferns and eighty wildflowers grow along the woodland trail. Designed to help visitors identify the plants, the trail has permanent markers with the names of both ferns and flowers. Many of the plants are rare or endangered, not commonly seen along the trails elsewhere. In the spring look for showy orchis, yellow lady's slipper, jack-in-the-pulpit, wild ginger, bloodroot, two varieties of hepatica, spring forget-me-not, May apple, trout lily, clintonia, and red trillium.

A sculpture garden adjoins the art center building, and towering above is a sugar maple estimated at over 250 years old. Admission is charged for the exhibits in the art center, but access to the trail is free. Contact the Southern Vermont Art Center, West Road, PO Box 617, Manchester, VT 05254; 802-362-1405.

## THE GARDENS AT HILDENE

Beautifully restored from original records, the formal gardens at this turn-of-the-century summer estate were first planted in 1907.

*Directions:* From Manchester Village, go south on US 7; the driveway for Hildene is on the left, just past the cemetery.

The summer home of Abraham Lincoln's son, Robert Todd Lincoln, Hildene was occupied by Lincoln descendants until 1975. The house overlooks the formal garden, and the best view of its design is from the second floor.

Fortunately for those gardeners who set about the restoration of these formal beds, Robert Lincoln was a man who kept everything, including meticulous records. Because of his penchant for storing even invoices for flowers, we know how many and what kinds of peony roots were ordered from France and that the initial plantings included 500 privets, 100 barberries, and 332 rose bushes.

The gardens were designed by Frederick Todd, who had been an apprentice to Frederick Law Olmsted, America's foremost garden designer of the time. The gardens, formal but set in parklike grounds, show the Olmsted influence. The beds are laid out in four squares, each surrounding an inner square of turf. At the far end, six beds form an arch, giving the design a graceful finish. Plantings were (and are) predominately privets, roses, and perennial flowers, with peonies still forming a showy display.

As interesting as the "upper gardens" on the terrace behind the house are the newly uncovered kitchen gardens that once stretched behind the impressive ninety-foot carriage barn. A potting shed, a greenhouse where annuals for the border plantings were grown, an herb garden, a grape arbor, and a terrace of brick paths separating beds of vegetables and cutting flowers once filled an area of several acres. A heavy lattice fence kept farm animals from straying into the vegetables.

The kitchen garden was more than a supply center. It was designed for quiet moments to appreciate its beauty as well. Down its center was a broad grass walk, in the center of which a brick path was later added. Wide flower beds lined the sides of the turf swath, which began at the greenhouse and ended in a well-kept woodland garden. Arbors covered in sweet-pea vines shaded wooden benches.

Slowly, these features are emerging once again—the brick path was covered in several inches of sod, and the grand allée of hawthorn trees had grown in so thick that passage through it was nearly impossible. A pond once filled with lilies was filled with brush, and most of the structures had long ago fallen down. But the remains were still there, and the gardens are slowly reappearing. Hardier plants survived: lilacs and peonies still bloom profusely although the eighteen-inch privets had become twenty-foot trees.

As we write, the restoration still has a long, slow way to go, but future plans include tours of the kitchen gardens and possibly even the sale of plants propagated from these historic specimens. If you are especially interested in the restoration of period gardens, the staff can, with sufficient prior notice, arrange a special tour while the work is in progress. Meanwhile, the rest of the gardens and grounds (which include an observatory) are well worth seeing. And, as their restoration progresses, the kitchen gardens will be one of the very few in existence from that era.

Tours of the house and access to the formal gardens and observatory are offered daily from mid-May through October; the grounds open at 9:30 A.M. and the last tour begins at 4:00 P.M. The gardens are at their best during June, July, and August. The property is owned by the Friends of Hildene, Box 377, Manchester, VT 05254; 802-362-1788.

## GRAFTON MUSEUM OF NATURAL HISTORY

In this small, but nicely arranged nature museum, the emphasis is on local flora, wildlife, and geology.

*Directions:* The museum is in the Town Hall in the center of the village, opposite the Grafton Inn. Grafton is west of Bellows Falls on State 121.

The exhibits in the Grafton Museum are well done, often with painted backdrops, such as the wetland scene behind the display of native ducks. A collection of winter weeds invites the visitor to match these to the list of plants; a darkened booth houses a fluorescent mineral "show." A working hive of bees attracts children to one corner, and a model of a local talc mine, complete with working railcars, tempts them to another. You can see dinosaur tracks, Indian artifacts, fungi, shells, corals, and the skulls of Vermont mammals, or you can learn the history of a tree from its rings and look inside a hornet's nest.

It is not a place you would plan to spend the entire afternoon, but the museum is certainly worth a stop if you are in Grafton when it is open. Special programs are announced from time to time, such as a trip to explore the ecology of a pond. A bulletin board on the Town Hall announces these events.

The museum is open from 1:00 P.M. until 4:00 P.M. on Saturdays and Sundays from Memorial Day through the end of October, as well as most other days during foliage season and occasionally by chance. Admission is free, but a container is there for donations. Contact the Grafton Museum of Natural History, Grafton, VT 05146; 802-843-2347.

## BROCKAWAY MILLS GORGE

The Williams River continues the process of riverbed erosion it began thousands of years ago when glacial Lake Hitchcock burst its restraints.

*Directions:* From Rockingham, follow State 103 west. About 2.5 miles beyond the Vermont Country Store is a road to the right signposted for the Madrigal Inn and Brockaway Mills. Follow that road as it winds along the precipice to the bottom of the hill where the road crosses a bridge and comes to a T. Park in the area to the right.

A few thousand years ago, when the great dam around Hartford, Connecticut, that held back the glacial meltwater of Lake Hitchcock was breached, it set in motion a process of water erosion of the Connecticut River Valley that is still at work. First, the water wore away at the main valley, where the river now flows. As it progressed, the wearing process spread up the valleys of the streams that fed the main river. The waterfalls on the Williams River shows the continuing effect of that phenomenon.

The gorge and its series of falls and cascades begin at the bridge and run east, after a right-angle turn, for almost 1,000 yards. Its bed is nearly eighty feet wide and, as it passes through the gorge, the river continues to sculpt large potholes and shape the rock. In spite of the fact that it is in a well-settled area, the gorge manages to retain its wild appearance.

Brockaway Mills is the fourth-largest gorge in the state. The falls are jagged and have a precipitous drop of about fifty feet over which the impounded waters above are allowed to fall. The best view is from the bridge, which looks over the shoulder of the falls into the chasm. Be careful here, however, because the bridge is narrow, the restraining cables weak, and passing cars perilously close. Downhill from the parking pullout, on the foundations of the old mill, you can get a more frontal view of the falls. Unfortunately, a power transmission station occupies the high ground that would give the optimum view. Another good view is from the Green Mountain Railroad excursion train as it crosses the gorge.

*Of interest nearby:* Perhaps the particular interest Vermonters take in studying the sky (see the entry for the Hartness Telescope) may result from the intense clarity of the stars in the vast areas far from city lights. It is a rare treat to be able to see stars from horizon to horizon, but that is possible on many Vermont hilltops. The Inn at High View has just such a location, and its lawns are a popular retreat for amateur astronomers, especially during the August meteor showers. In the spring, its gardens and woodland wildflower paths are an added attraction. To reserve your piece of sky, contact the Inn at High View, RR 1, Box 201A, Andover, VT 05143; 802-875-2724.

*Also of interest:* For those who enjoy walking along woodland roads, Rowell's Inn in nearby Chester is one of a small group of inns that offers a program of walking from inn to inn. Or, the owners of Rowell's will take guests to Weston, 5.5 miles away, and leave them there with a map and description of the route back. The slow pace of walking leaves plenty of time to savor the views and investigate the flowers and birds in this thinly settled area, as well as whet your appetite for the fine dinner awaiting

---

### VERMONT'S RIVERBEDS: PERSISTENCE PAYS

One of the outstanding things about the state's roads, aside from their general good condition, is the fact that they travel alongside many of the streams that drain its hills and mountains. These streams are literally the result of thousands of years of erosion by enough water to probably more than fill the Atlantic. In times of tremendous flood and drought, they have continued flowing, wearing away the softer rock, eating away at the dams of upthrust rock, and carrying huge boulders downstream to smash at the bed and sides of the stream, making gravels and sands. Each of these stream beds has its own story written in the rock walls of its sides and in the boulders that line it. Take time to examine the differences in the rocks that lie there. The pieces of black slate, small rounded pieces of quartz, or rounded pieces of granite all came from different places. Compare them with the rock in the bed or on the walls of the stream and see if they match or were carried there from miles away.

---

you at Rowell's. For information, contact Rowell's Inn, RR 1, Box 267-D, Chester, VT 05143; 802-875-3658.

## OLD JELLY MILL FALLS

More than a cascade, these falls are really a series of small waterfalls that form a stair.

*Directions:* Take State 30 northwest from Brattleboro about 3.5 miles to Stickney Brook Road. Just beyond the road and bridge is a place to park on the west side of the road. You can also park on the side of Stickney Brook Road, but it is narrow and space is limited. The falls are just off the road.

A dirt track leaves from the rear of the parking area and immediately goes to the brook, which is only a few minutes' walk. This is a cool and shaded place where one longs to linger on hot summer days. Stickney Brook is a broad, shallow stream passing through a continuous arch of tall trees, in a string of falls covering a distance of more than 100 feet. The water drops in sheets over a number of stairlike short falls, forming shallow wading pools in the stream bed. Along the side of the stream are the stone foundation walls of the old mill that gave the falls their name.

## DUMMERSTON GRANITE QUARRY

Granite quarrying was a major industry in northern New England during the nineteenth and early twentieth centuries, and this abandoned site is a poignant reminder of those days.

*Directions:* On State 30, a very short distance north of Stickney Brook Road (see Old Jelly Mill Falls), take a right across a green steel bridge. On the other side, go left onto Quarry Road. A very short distance up the road you will see the remains on the right side of the road.

The first sign of the quarry is a crumbling building with a huge power shovel rusting away. A few feet beyond sits an old trailer and scattered about the site are other large pieces of abandoned machinery with large trees growing up through them. To the rear of the site is a large wall of granite blocks that supported the works and across the road is a hill of tailings. A few feet farther on you can see the quarry through the young forest. The bottom of the quarry is filled with greenish blue water, and the fractured hillside displays graffiti painted by youths braver than they were smart. There are signs posting the property, but the most interesting parts of it are visible from along the side of the road.

## FISHER-SCOTT MEMORIAL PINES

The Vermont author Dorothy Canfield Fisher found inspiration in these woods, which have been named a National Natural Landmark for their size and age.

*Directions:* From US 7A, north of Arlington, take Red Mountain Road to the west. It leaves the main road .4 mile north of the bridge over the Battenkill River. In .2 mile, just before the pavement ends, you will see a trail on your left and a narrow pullout space, which is the entrance to the woods. Just inside the woods, a plaque on a stone notes the designation as a Natural Landmark.

The old-growth pines here are among the tallest and oldest in Vermont, and by some estimates, they are *the* largest white pines in the state. Heights reach 120 feet, and diameters are as great as 40 inches. These white pines are not a climax forest, but the first stage in a succession forest. Even now, younger hardwoods are growing among them, and as the older trees eventually die, the hardwoods will have more light and will take over as the predominant tree species. These pines are thought to be about 185 years old.

A short trail leads into the woods, where, like Dorothy Canfield Fisher, you may find inspiration. Follow it through these giants and you will come to a second marker on a rock, a poignant memorial to three generations of the Fisher family. Directly behind it stands the largest of the trees.

## HEALING SPRINGS NATURE TRAIL

The water of these springs was once bottled and sold for its healing qualities, but the springs were eventually dammed to create power to drive a mill. When the mill closed, the lake was maintained as a colony of cottages until the state acquired it.

*Directions:* The nature trail is at Shaftsbury Lake, north of Shaftsbury on US 7A. A Shaftsbury State Park sign marks the entrance on the east side of the road. Park in the lot at the end of the road and walk back across the base of the dam to the beginning of the trail.

On your way to the beginning of the trail on the west end of the dam, notice the small mineral spring at the foot of the dam. It contains organisms that are peculiar to that environment. The nature trail is an easy walk and should take a half hour to forty-five minutes, depending on your pace. The trail illustrates not only the effect of man upon the land, but the effects of the last glacial period. As you skirt the lake, you will see the cottages and Scotch and Norway pines, non-native trees planted when they were built. The trail continues on a small bridge over a marshy bay onto a long low hill. This is an esker, an accumulation of gravel and stones that gathered in a tunnel through the overlying glacier as it melted. When the glacier completely melted away, the bed of the stream moved elsewhere and this causeway of gravel was left high and dry. Look off the sides through the beech and hemlock forest and note the difference between the marshlands on one side and the lakeside on the other. Few places illustrate the shape and composition of an esker as well as it is shown here.

The trail then passes down to lake level and through a low marshy area, partly over a boardwalk. Fine examples of marshy and wetlands plants are found here, including several varieties of ferns, red-osier dogwood, speckled alder, cattails, and bitter-sweet nightshade. An abandoned rail line, built in 1856, cuts across the esker and forms an elevated line along the eastern side of the park. Huge "cabbage" pines, so called because of their spreading shape, indicate that this area was an open field before the building of the railroad.

The trail shows the effect of humans on the environment, but equally powerful is its demonstration of the power of nature to overcome or adapt to the situations created by humans. An excellent brochure describing the history and environment of the area is available at the park. The park has a day-use area that includes swimming, changing rooms, and picnic areas with grills.

## PARK-McCULLOUGH HOUSE GARDENS

Gardens dating from the mid-nineteenth century have been restored and maintained on the grounds of an impressive Victorian home, also open to visitors.

*Directions:* From US 7 or 7A follow State 67A west to North Bennington. The Park-McCullough House is signposted from State 67.

Several gardens surround the huge ornate mansion and its carriage barns. The showiest is the perennial garden, privately owned and in the care of a retired botanist who has created beds of flowers that seem to bloom forever. Giant black-eyed Susans, bee balm that looks you straight in the eye, pink mallow, frothy white hydrangea, phlox, achillea, and lilies in shades of orange, red, gold, and pink, all in full bloom at once.

In addition, there are herb beds, a rock garden, and a community garden with beds raised for handicapped access.

A cast-iron framed grapery has recently been restored. Once a year, in early June, plants from the garden collections are available for sale.

The grounds are open at all times, free of charge; the best months for visiting the gardens are June, July, and August. Tours of the house are offered (admission is charged) from late May through October, hourly from 10:00 A.M. until 3:00 P.M. Contact the Park-McCullough House, PO Box 366, North Bennington, VT 05257; 802-442-5441.

## BENNINGTON ROAD CUT

The building of State 9 required cutting through a ridge, exposing mineral specimens, which can be collected easily.

*Directions:* Traveling west on State 9, the first of the two cuts is 4.7 miles west of the entrance to Woodford State Park. A parking pullout is just beyond the cut. The second is 3.6 miles farther west, 2.5 miles beyond the Long Trail crossing. From the west, the first cut is 2.6 miles east of the junction with US 7 in downtown Bennington. The second is 1.1 miles from the Long Trail crossing.

The easternmost of the cuts is known to rock hounds for its blue quartz, which is of jewelry quality. Garnet, hornblende, and orthoclase are also found here. The western cut is of Cambrian Cheshire quartzite, a pink stone whose color is quite striking even in passing. Since the cuts are in the Green Mountain National Forest, collectors are allowed to pick up any loose pieces, but cannot use a pick, hammer, or other tool to dislodge specimens. Call the Forest Service with any questions, at 802-362-2307.

## LUMAN NELSON MUSEUM
## OF NEW ENGLAND WILDLIFE

A rather solitary Yankee's lifelong fascination with wildlife stays alive in this fine collection of specimens of native mammals and birds, some of which are now extinct.

*Directions:* From Brattleboro, or Bennington, take State 9 to the top of Hogback Mountain. Park at the scenic overlook. For times, call 802-464-5494.

Luman Nelson was born in 1874, and he lived most of his life across the river in Winchester, New Hampshire. Throughout his life he had a fascination with animals, particularly those in the woods around him. This led him to learn taxidermy, and by the time of his death he had preserved hundreds of examples, including some species that have become rare or extinct since he mounted them. In a classic case of misguided frugality, the State of New Hampshire turned down a chance to acquire the collection when he died and it was bought by private individuals who have assembled it for public view.

One doesn't expect to find a collection of this importance in a private museum located in a highly touristed gift shop. The collection covers two floors of space that most businesses would put to a more commercial use. It includes rarities such as albino deer, squirrels, and red foxes. There are several bobcats, long disappeared from most of New England, and the more common species, such as woodchuck. The lower level has an outstanding collection of birds, including wild turkey, about every variety of duck that ever set its webbed feet on these shores, as well as geese, falcons, hawks, and eagles. One of the true wonders of Vermont is the variety of its wildlife, and this is a good place to meet and study it. The displays aren't fancy, but they are well lighted and easy to see.

**Red foxes eat almost anything: insects, berries, rabbits, and mice**

## VERNON BLACK GUM SWAMP

A relic of a forest of 4,000 years ago, these swamps contain fine examples of black gum, a tree that survives only several zones south of Vermont but at this latitude only in the special conditions that exist in this spot.

***Directions:*** State 142 follows the Connecticut River south from its intersection with State 119 on the southern end of the Brattleboro business district. Take it to the Vernon town clerk's office in the modern town buildings. In order to protect this sensitive site, the Town of Vernon requests that you call the town clerk's office (802-257-0292) during the week. They will give you directions. The site is easy to find and readily reached by auto and a short hike of about half a mile for the high swamp trail and 1.2 miles for the lower swamp trail. A trail map is available.

The Black Gum Swamp is situated in the J. Maynard Miller Town Forest. We had the personal pleasure of touring the swamp with Mr. Miller, who is himself one of the natural wonders of Vermont. A selectman for over fifty years and a lifelong farmer, he was the force behind the preservation of this special place for future generations, an example of common-sense preservationism.

From the parking area, follow the signs for the Black Gum Swamp on the western end along a road and past a house to another small parking area. The swamp trails depart from the end of that lot, the high swamp following red markers and the low swamp following silver markers. Two other trails are also available, the overlook (green markers) and the laurel trail (blue markers).

The swamps are a complex ecosystem that evolved following the tertiary period when the warming of the earth allowed southern species to expand their habitat into the north. When global cooling began about 3,000 years ago, the viability zones of those species retreated, but unique pockets remained. Growing in kettle holes left behind when huge buried pieces of the last glacier melted, these special pieces of forest have never been harvested.

The black gum, while common in the South, is one of the strangest trees in the northern forest. Their tall, thick bodies rise to thin, sparse tops that seem to have been severely trimmed. The oldest trees in these forests are slightly larger than two feet in diameter and have been estimated to be about 400 years old. They are easily identified because the tall ones tend to lean slightly, and while the bark on the upper side of the lean is relatively thin and smooth, the bark on the lower side is thick and very deeply fissured, up to five inches deep. The oldest of the trees are still alive, even though they may be completely hollow with rot or nearly fallen to the ground.

While there are many specimens of the black gum, you'll also encounter fine examples of other species in the swamp, including red maple, oak, hemlock, and beech. Covering the floor is a rich deep layer of sphagnum moss that is so thick and luxuriant between the trees and bushes that it seems to be a carefully tended lawn. Hobblebush (one of the main supporting plants of deer), witch hazel, winterberry holly, mountain holly, and high-bush blueberry also grow in abundance.

Mountain laurel and sheep laurel grow in the swamp and along the access trails. These, and the black gum, bloom in June. Wildflowers include pink lady's slipper, goldthread, and Indian pipe. Several fern varieties can be found, including cinnamon ferns four to five feet tall, royal ferns, and Virginia chain ferns. On the north end of the high swamp, a large rock outcrop has mosses and tiny polypodia ferns growing from the surface among mosses and tiny hemlocks.

In addition to the swamps, this site has a large white birch and laurel grove on the green (Overlook) trail, a notable destination in its own right. While there, watch for the few remaining

**Although Indian pipes lack chlorophyll in their leaves, they are
true plants, not fungus, which get their food from close association
with a fungus beneath the soil**

examples of American chestnut. These trees, growing largely
from root shoots of dead parents, don't reach the splendid
proportions of their ancestors but give hope that the species will
rebound from its state of near extinction. On this trail is a clearing with a fine view of Mount Monadnock, forty miles away in
New Hampshire. It's a prime spot for a picnic.

## VERNON FISH LADDER

While misuse of the Connecticut River has caused the virtual disappearance of the salmon and shad populations that once filled it, this $10 million project seeks to secure their return.

*Directions:* Take State 142 south from Brattleboro and at the town center take Governor Hunt Road (follow the signs to Vermont Yankee) passing by Vermont Yankee and the hydroelectric plant to the parking area on the southern end of the dam. There are signs at the site.

A six-panel display in a gazebo on the access road explains the power plant, the mechanics of the fish ladder, and the importance of the restoration of the river for the fish species found here. The 984-foot-long fish ladder contains 51 steps, 26 of which raise the fish one foot each, with the remainder raising them in six-inch increments, for a total rise of over 40 feet. A special viewing window permits visitors to see the fish as they work their way up the ladder.

The ladder is open during the salmon and shad runs mid-May to mid-July and occasionally during the fall. For information, call the Superintendent of the Vernon Station at 802-254-4388 during the week.

Just south of the ladder is Governor Hunt Park, which, in addition to picnic facilities, has a boat and canoe put-in below the dam. This is also the portage put-in. It's open 6:00 A.M. to 9:30 P.M. Use the first drive for the picnic area, the second for the put-in.

## WHERE TO SEE FALL FOLIAGE

By far the best-known of all Vermont's natural wonders is the brilliant and colorful display of foliage each autumn. While a

### AUTUMN SPLENDOR: WHY LEAVES CHANGE COLOR
For about one month a year the deciduous trees of Vermont turn colors in shades that range from pale yellow, through all oranges, and into reds that sometimes come close to purple. Why would a tree shed a perfectly good leaf, and above all why would it change its color? The answer lies in the fact that as the sunlight changes angle, intensity, and duration in the fall, the trees prepare for winter by shutting down their production of chlorophyll. When they do this, the other chemicals that have been there all along begin to show. Two of these, xanthophyll (what makes eggs yellow) and carotene (what gives carrots color), make the leaves turn yellow. Another chemical, anthocyanin, is the agent that makes them turn red. Different combinations of these chemicals in any given tree is what causes the varying shades of orange. There is even one tree that looks like an evergreen variety that joins in this display. The fine leaves of the tamarack, or larch, a tree that likes low wet areas, also turn to yellow and then brown as the season advances. During September and October, this change starts first in the highlands and in the north and spreads south, the best of the color staying in any one place for about two weeks.

drive along any road in the state will offer a good view of this annual spectacular, everyone has his or her favorite routes. Since the southern roads lead from our back door, it is natural that our personal preferences—those roads we drive each year just to satisfy our souls—are those in southern Vermont.

Beginning in Brattleboro, go north on State 30 past Maple Valley Ski Area 1.4 miles, taking a left onto the unnumbered road to West Dummerston. Continue through South Newfane and East Dover to West Dover, where you will join State 100. Go north to West Wardsboro and take the unnumbered road west marked for Stratton. It will become a good dirt road and take you across the spine of the Green Mountains to a settlement called Kansas. Go left over the bridge through East Arlington and left again at the

T to Arlington. In East Arlington, take a short detour north to Chiselville to cross the covered bridge. Park on the far side and walk back to look down into a deep gorge. One wonders how they built the first bridge there.

Another side trip from Arlington is west on State 313 to the New York border and back, along the valley of the Battenkill. Returning to Arlington, take US 7A south to Bennington and return to Brattleboro on State 9, stopping for the views on top of Hogback. If you are feeling adventuresome, go south on State 8 in Searsburg, bear left onto State 100 through Readsboro to Jacksonville, then try to find your way eastward on the maze of dirt roads to West Guilford or Green River (a fine little town with a covered bridge in the center and no paved roads in any direction) and on US 5, then north to Brattleboro. If you can't find your way through, one of two things will happen: You'll veer northward and come out on State 9 again, or you'll see a sign telling you that you've crossed into Massachusetts. In the latter case, just turn around and go back to the next intersection and go right. You might have to do this a couple of times before you get to Green River, but you can't get too far lost, and with every tree ablaze with color, it doesn't really matter which road you are on.

This route is just as easily started in Bennington or Arlington, but you should try to time it so you are not traveling west late in the afternoon, to avoid having the sun in your eyes. A shorter adventurous trip is to leave Brattleboro on State 9 and go as far as Searsburg, then go south, following the same route (or nonroute) as above back to Brattleboro.

For a view from higher up, with more open vistas, leave Brattleboro on I-91 and go north to Springfield. Take State 11 to Chester (stopping for tea and scones at the Rose Arbor just off the green on School Street), then go south on State 35 to Grafton, south to Townsend, and south on State 30 through Newfane and back to Brattleboro.

A nice back-road route from Arlington begins by traveling north on US 7A to Manchester Village, then taking West Street across to State 30 (cutting off the traffic tie-ups in Manchester Center). Go north. If the traffic is heavy, you can bypass Dorset by taking the parallel road to the west of State 30 in South Dorset to East Rupert. Whichever route you take there, go west on State 315 through Rupert to West Rupert. Take the unnumbered dirt road south in West Rupert (this road is not for wimps, but is quite passable), going left at the T and coming out just south of Sandgate. Take a short detour to the left and up the hill to the Sandgate town office building for a view of the high mountain pastures, then turn around and head south to State 313. A left turn will bring you back into Arlington.

# Index